Tassiane Bolzan Morais
Alexandre Swarowsky

Efficiency of nitrogen doses and shading in tomato cultivation

Tassiane Bolzan Morais
Alexandre Swarowsky

Efficiency of nitrogen doses and shading in tomato cultivation

In protected cultivation

ScienciaScripts

Imprint
Any brand names and product names mentioned in this book are subject to trademark, brand or patent protection and are trademarks or registered trademarks of their respective holders. The use of brand names, product names, common names, trade names, product descriptions etc. even without a particular marking in this work is in no way to be construed to mean that such names may be regarded as unrestricted in respect of trademark and brand protection legislation and could thus be used by anyone.

Cover image: www.ingimage.com

This book is a translation from the original published under ISBN 978-613-9-68880-7.

Publisher:
Sciencia Scripts
is a trademark of
Dodo Books Indian Ocean Ltd. and OmniScriptum S.R.L publishing group

120 High Road, East Finchley, London, N2 9ED, United Kingdom
Str. Armeneasca 28/1, office 1, Chisinau MD-2012, Republic of Moldova, Europe
Printed at: see last page
ISBN: 978-620-8-23167-5

CONTENTS

1 INTRODUCTION

Tomato fruits are part of the diet all over the world (GUIL- GUERRERO & REBOOLLOSO-FUENTES, 2009). In Brazil, an area of 55,000 hectares was cultivated in 2015, with the states of Sao Paulo, Goiás and Minas Gerais being the largest national producers (MAPA, 2015). In 2015, the Brazilian tomato crop, including the two segments: industry and table tomatoes, totaled 4,145,553 tons of fruit (IBGE, 2016). As a result, tomatoes are a very important vegetable on the national market and are part of the diet of a large part of the Brazilian population on an almost daily basis.

Due to the significant increase in consumer demand for tomatoes, it is becoming increasingly necessary to maintain and increase the quality of their fruit and also to maintain and increase productivity. Consumers are increasingly demanding high quality fruit (IGLESIAS et al., 2015). Tomato consumption is currently linked to a nutritional indicator of good eating habits and quality of life (HERNANDEZ SUAREZ et al., 2007). The quality demanded by consumers, especially in horticultural crops, is related both to aspects of the appearance, texture and taste of the fruit which arouse consumer interest, and to its nutritional value (IGLESIAS et al., 2015), qualities which are associated with characteristics such as soluble solids content, pH and titratable acidity (DING et al., 2016).

Several factors influence the final yield and nutritional quality of tomato fruit. Among the factors that influence the growth and production of tomato plants is the adequate supply of nutrients from the soil to the plants, especially nitrogen (FERREIRA et al., 2010). According to Araùjo et al. (2007), in Brazil, nitrogen fertilization for tomato plants is recommended empirically and is rarely based on relationships derived from recommended doses and the expected yield of the crop.

In relation to the application of nitrogen and the quality of tomato fruit, the nutritional quality of the fruit is influenced by the nutrient nitrogen through its availability, for example, by moderately reducing the supply of nitrogen, a better sugar content of the fruit can occur without affecting the productivity of the crop, and a reduction in the supply of nitrogen combined with an increase in the radiation on the fruit can be used to improve the quality of the fruit through an increase in the antioxidant content of the fruit (BÉRNARD et al., 2009). An adequate supply of N during the growth and development of the tomato crop guarantees high productivity (high yields) and maximizes profits for the producer (ELIA & CONVERSA, 2012). However, it is worth noting that the response of a crop, such as tomatoes, to the availability of nutrients can vary according to cultivars, external factors, cultural practices, substrates and environmental conditions (PASSAM et al., 2007).

In addition to nutritional deficiencies, horticultural crops, especially tomatoes, have production problems related to bad weather. Therefore, constant technological innovations to overcome these problems must be studied. To minimize the negative effects of the climate on tomato crops, one solution has been to use shading screens over the crops and to grow them in a protected environment. In addition to the climatic effects, crops grown in a protected environment allow producers to plan their production, set and meet productivity targets and also meet deadlines for delivering the product to the consumer (PAIVA, 1998).

Due to the relative ease of management of crops grown in protected environments when compared to conventional open field systems in Brazil, growing vegetables in these conditions has been gaining ground among producers (CARRIJO et al. 2004). Growing tomatoes in a protected environment, seeking a change in the microclimate of this environment, aims to solve problems related to the agricultural production of vegetables at times that are less favorable to their cultivation (MARTINS et al., 1994). This is extremely important as it helps to regularize the supply of the crop, obtain higher prices and production in the off-season (STRECK et al., 1998).

In order to reduce the damage caused by adverse weather conditions, especially radiation and temperature, shading screens have been used on many agricultural crops, including tomatoes (SILVA et al., 2013). According to Rjapakse & Shahak (2007), in addition to weather damage, shading screens protect against damage caused by rain, hail and wind and also protect crops from birds and insects.

Thus, the study of nitrogen fertilization and shading in tomato cultivation under a protected environment and the interaction between these two factors is indispensable for obtaining adequate information on these management practices in order to maintain the crop's production in the off-season, its yield and fruit quality, guaranteeing income for the producer and possibly reducing the use of nitrogen.

2 OBJECTIVES

2.1 GENERAL OBJECTIVE

The general objective of this work was to determine the influence of nitrogen fertilization, shading and the interaction between them on the agronomic performance of tomato crops in a protected environment.

2.2 SPECIFIC OBJECTIVES

To determine the performance of the tomato crop when exposed to different doses of nitrogen in a protected environment;

To evaluate the influence of using a shading screen on the tomato crop in a protected environment;

To verify the interaction of the factors evaluated, nitrogen doses and shading screen, on the agronomic performance, productivity and fruit quality of the tomato crop in a protected environment.

3 LITERATURE REVIEW

3.1 GENERAL APPEARANCE OF TOMATO PLANTS

3.1.1 Origin and history of the tomato

The tomato (*Solanum Iycopersicum* L.) originates from the wild cherry tomato (*Solanum lycopersicum* var. Cerasiforme), which comes from a territory bordered to the north by Ecuador, to the south by northern Chile, to the west by the Pacific Ocean and to the east by the Andes Mountains. Before Spanish colonization, the tomato was taken to Mexico where it was cultivated and improved (FILGUEIRA, 2008). The tomato was introduced to Europe through Spain between 1523 and 1554 for ornamental purposes, as it was thought at the time that the fruit was inedible (FILGUEIRA, 2008).

There is evidence that the Italians were the first to cultivate the tomato fruit, around 1550, initially out of curiosity for the crop and the ornamental value of its fruit (FILGUEIRA, 2000). The production and consumption of the fruit spread to the United States *in* the 19th century and, to this day, *in* addition to its *fresh* fruit, its derived products, such as soups, sauces, drinks and ketchup are consumed regularly (HARVEY et al., 2002).

In Brazil, the habit of eating this fruit was introduced by European immigrants at the end of the 19th century. Today, the crop is grown all over Brazil and the world. Tomato cultivation became important worldwide in the 1900s and is among the three most industrialized foods, along with garlic and onions (FILGUEIRA, 2000).

3.1.2 Importance and production of tomato growing in Brazil and worldwide

Tomatoes are considered to be the most industrialized vegetable in the world and are therefore of great socio-economic importance (KROSS et al., 2001). Tomato cultivation is divided into two production chains: the table segment, where the fruit is consumed *fresh;* and the industrial segment, where the fruit is processed. Each of the production chains has different characteristics, from management and production (cultivars, cultivation methods and final consumption) to the processing, processing and marketing of the fruit (SANTOS, 2009).

According to data from FAOSTAT (2013-a), the world's largest tomato producers are: China, the United States, India, Turkey, Egypt, Italy, Iraq, Spain, Brazil and Mexico, respectively. Together, these countries produce 76% of the world's tomato crop. In recent years, there has been a considerable increase in tomato production. In 1992, world production was only 74.9 million tons; in 2012, production exceeded 161 million tons. In percentage terms, this is an increase of 116% in the last 20 years. World tomato consumption went from 12.6 kg per person per year in 1989 to 20.5 kg per person per year in 2009 (FAOSTAT, 2013-b).

Brazil is the ninth largest tomato producer in the world. Data from FAOSTAT (2013-a) shows that in the last twenty years, between 1983 and 2013, Brazilian production went from 1,550,778 tons to 4,187,646 tons. And in the last six years, Brazil had a yield of 634,958 kg ha^{-1} in 2008 and 668,025 kg ha^{-1} in 2013. With regard to the Brazilian states, data from IBGE (2016) for the 2015 harvest shows the state of Sao Paulo as the largest national tomato producer, contributing with a planted area of 14,967 ha, followed by the states of Goiás and Minas Gerais. The state of Rio Grande do Sul is in ninth place.

Production costs for tomatoes average between R$25.00 and R$30.00 per 23 kg box, and are sold at an average of R$120.00/box (CEPEA, 2016).

3.1.3 Botanical and agronomic aspects of the tomato plant

The tomato plant belongs to the Solanaceae family and can grow in a creeping, semi-erect or erect form. Growth can be limited in determinate cultivars and unlimited in indeterminate cultivars, the latter of which can reach 10 m in one year. They can grow in a range of latitudes, soil types, temperatures and cultivation methods. Warm environments with good lighting and drainage are the most suitable for their cultivation (ALVARENGA, 2013).

Tomatoes have a large root system, consisting of a main root with a large number of secondary roots and a large number of adventitious roots growing from the base of the stems. The stem is angular, covered in perfectly visible hairs, many of which are glandular in nature, giving the plant a characteristic smell (ARGERICH & TROILO, 2011).

The flowers of the tomato plant are considered perfect because they have an androecium and gynoecium and their stylus is protected by a cone of five or six anthers. The species *S. lycopersicum*, considered the tomato species of agricultural interest, has six anthers; the others only have five. The ends of the anthers are tapered, devoid of pollen and the tube narrows. The ovary can have two, three or four locules, or be multilocular, and is usually self-pollinated, with a low incidence of fruit from crosses (GIORDANO & RIBEIRO, 2000).

Tomato fruits are fleshy, juicy berries that vary in appearance, size and fresh mass, depending on the cultivar. When ripe, they have a reddish color, resulting from the combination of the color of the flesh and the yellow outer skin. There are some exceptions, such as the Japanese cultivars of the "salad" group, which are pink due to their whitish skin. The fresh mass of the fruit varies widely, from 25 g ("cherry" type) to 400 g ("salad" type) and the seeds are hairy, small and when present inside the fruit they are surrounded by mucilage (FILGUEIRA, 2008).

The tomato crop cycle can be divided into three phases. The first phase is between 25 and 35 days after sowing until the start of flowering. The second phase begins when the crop blooms and ends when the fruit is harvested. Finally, the third phase covers the entire harvest (ALVARENGA, 2004).

3.2 TOMATO FERTILIZATION

3.2.1 Mineral fertilization

Mineral nutrients are elements obtained from the soil, especially in the form of inorganic ions. Although nutrients circulate through all organisms, they enter the biosphere through plant root systems and are translocated to the various parts of plants, from where they are used for numerous biological functions (TAIZ & ZEIGER, 2009). According to the same author, high agricultural yields depend on fertilization with mineral nutrients.

Nutrients are classified into three distinct groups according to their redistribution within plants: the mobile ones ($NO3^-$, $NH4^+$, P, K and Mg), those considered to be moderately mobile (S, Mn, Fe, Zn, Cu and Mo) and the immobile ones (Ca and B). The immobility of a nutrient is associated with a lack of mobility in the phloem, preventing its remobilization (EPSTEIN & BLOOM, 2005).

For plant growth, there are some elements classified as essential elements, defined as those that are intrinsic components in the structure or metabolism of a plant or whose absence causes severe abnormalities in the plant's growth, development and reproduction (EPSTEIN & BLOOM, 2005). There are two key factors for plants to achieve a balanced metabolism, high dry matter production and adequate development: sufficient quantities of nutrients and their proportions being balanced (LARCHER, 2000). When the supply of nutrients to plants is inadequate, physiological disorders occur which manifest themselves through symptoms of deficiency or toxicity. These symptoms will depend on the function that the nutrient performs for the plant and its capacity for mobility in the phloem and xylem (LARCHER, 2000).

3.2.3 Assimilation of nitrogen from the soil and absorption by plants

Nutrient assimilation is the process by which nutrients obtained by plants are incorporated into carbon compounds necessary for their growth and development. These processes involve highly energetic chemical reactions (TAIZ & ZEIGER, 2009).

Among the elements considered essential for plant development is nitrogen (N). Of the nutrients that are absorbed from the soil, nitrogen is the one most agricultural crops require, and its essentiality is related to the constitution of proteins, amino acids, pigments, nucleic acids, hormones, coenzymes, vitamins and alkaloids (FLOSS, 2011).

Nitrogen is present in many forms in the biosphere. However, most of it is unavailable to living organisms in organic form. To obtain this nitrogen, reactions known as nitrogen fixation must take place, either by industrial or natural processes, to produce ammonium (NH3) or nitrate ($NO3^-$) - sources of inorganic nitrogen (Figure 1). When fixed into ammonium or nitrate, nitrogen enters the biogeochemical cycle, passing through organic or inorganic forms before returning to the form of molecular nitrogen. The ammonium ($NH4^{+}$) and nitrate ($NO3^{-}$) ions, generated by fixation or released by the decomposition of soil organic matter, become objects of competition between plants and microorganisms (TAIZ & ZEIGER, 2009).

Figure 1: Nitrogen cycle. Source: Adapted from Taiz & Zeiger, 2009.

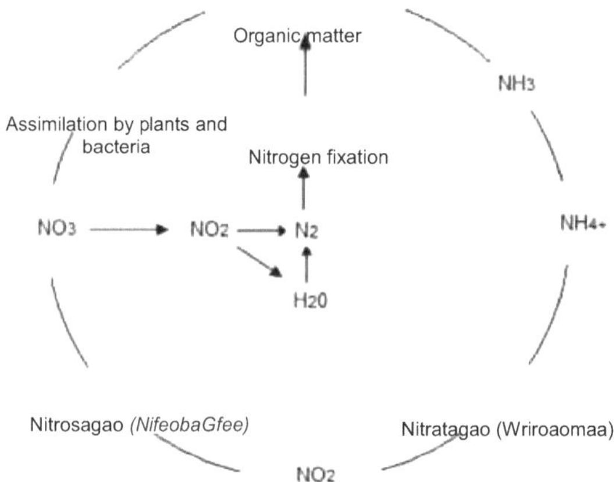

Plant roots actively absorb nitrate from the soil solution through nitrate-proton co-transporters (CRAWFORD & FORDE, 2002). However, plants assimilate most of the nitrate in organic compounds, with the reduction of nitrate to nitrite being the first stage in the process (OAKS, 1994). Nitrite ($NO2^-$) is highly reactive and toxic. When plant cells transport nitrite into the chloroplasts in the leaves and the plastids in the roots, nitrite is reduced to ammonium by the enzyme nitrite reductase (TAIZ & ZEIGER, 2009). When plant roots receive small amounts of nitrate, it is reduced in these organs, but when the supply increases, a proportion of the nitrate is translocated to the aerial parts, where it is assimilated (MARSCHNER, 1995).

The ammonium cation is a positively charged polyatomic ion with the chemical formula NH4+. It is derived from root absorption or produced by nitrate assimilation or photorespiration, being converted into glutamine and glutamate, which are located in the cytosol and plastids of the roots or chloroplasts. Once assimilated, nitrogen can be transferred to other organic compounds through various reactions (TAIZ & ZEIGER, 2009).

As ammonium is an easily reacted compound, it can be lost from the system through volatilization. The loss of ammonium is caused by the deprotonation of the NH4- ion, converting it into ammonia (NH3), a highly volatile compound which is then dissipated into the atmosphere. These losses that occur in soils depend on their pH, and in acidic pH conditions the predominant chemical species is NH_4^+. In alkaline soils or soils with a pH greater than 7, any nitrogen fertilizer containing ammoniacal nitrogen is subject to NH_3 losses through volatilization, but the occurrence of soils with these characteristics in Brazil is very low (TROEH & THOMPSON, 2007).

The appropriate choice of nitrogen source must be taken into account, since plants respond differently to different forms of N, but we know that it is preferentially absorbed by the roots in inorganic form, such as nitrate (NO_3^-) or ammonium (NH_4^+) (GHANEM et al.; 2011; MARTÎNEZ-NADAJÙR et al., 2013). The absorption of NH_4^+ requires less metabolic expenditure than the absorption of NO_3^-, since the latter, when absorbed, needs to be reduced in order to be assimilated (BRITOO et al., 2001).

Plants absorb practically all mineral nutrients from the soil solution. For a nutrient to be absorbed by the root, there must be contact. The vast majority of nitrogen is absorbed through mass flow, where the nutrient moves through the soil due to a difference in potential from a wetter region (higher potential) to a drier one (lower potential) (FLOSS, 2011). This difference can be generated by the absorption of water by the plant, allowing a nutrient that is far from the root to move to the root (FLOSS, 2011). Nitrate (NO_3^-) and ammonium (NH_4^+) pass through the plasma membrane (plasmalemma) of the epidermal cells and the root cortex via specific transporters for these forms of nitrogen (LARSSON & INGEMATSSON, 1989).

The amount of nitrogen to be absorbed varies during the plant's development cycle depending on the number of roots and the rate of absorption. Normally, this amount increases progressively during the vegetative growth period, reaches a maximum during the reproductive stages and falls at the stage of grain or fruit filling (CREGAN & BERKUM, 1984).

3.2.2 The influence of nitrogen on tomato cultivation

In addition to other factors, the growth and development of the tomato crop depends on an adequate supply of nutrients. Taking into account the plant's physiological processes, N has the greatest influence on growth and absorption of the other nutrients, and is considered the most important in terms of controlling optimum crop nutrition (HUETT & DETTMANN, 1988).

Nitrogen (N) occupies a prominent position in tomato cultivation in terms of the quantities and sources required. With regard to quantities, it has been considered that a ratio of 2.0 to 2.5 g of N for each kilogram of fruit obtained is satisfactory (SCAIFE & BAR-YOSEF, 1995). Nitrogen sources (nitric, ammoniacal, amidic) do not significantly affect the production and quality of tomato fruit; however, among these, urea has the highest financial return for tomato plants (ALMEIDA, 2011). With regard to the amount of the nutrient extracted by tomato plants, every ton of fruit harvested contains 3 kg of nitrogen (EMBRAPA, 2003). The fruit is a major drain on nutrients and photoassimilates, and these nutrients, especially nitrogen, are exported along with the fruit (ARAÙJO, 2003).

Evaluating the levels of nitrogen applied to tomato plants, with the following installments: the entire dose of N applied at planting; one third of the N applied at planting and two thirds at 25 days; one third of the N applied at planting, one third at 25 days and one third at 50 days; half of the N applied at 25 days and the other half at 50 days after planting, the authors concluded that among the installments of N doses, the most efficient is the one applied in three doses throughout the crop cycle (FARIA et al., 1996).

In tomato growing, increasing the dose of N to the point where it can be absorbed without causing phytotoxicity in the plant contributes to greater plant growth, an increase in the dry matter of the roots, stem, leaves and fruit, an increase in plant height, the number of leaves and leaf area, an increase in the quality of flowering, fruiting and productivity (FERREIRA et al., 2010). Thus, adjusting the nitrogen fertilizer recommendation is extremely necessary in order to obtain high yields from tomato plants and maximum economic return from the activity (ARAÙJO et al., 2007).

Some external factors such as irrigation, rainfall, how the fertilizer is applied, the organic matter content in the soil, the crop that preceded the cultivation, the original nitrogen content in the soil and the production potential of the tomato crop in the specific production system can influence and make it difficult to correctly quantify the dose of nitrogen to be applied (ARAÙJO et al., 2007).

With regard to the quality of tomato fruit, several authors have researched the effect of increasing or decreasing the availability of N to the plants. Some characteristics must be taken into account when determining fruit quality, such as pH, soluble solids content, titratable acidity, color and fresh weight (ANAÇ et al., 1994). These characteristics can be directly affected by nitrogen fertilization, as well as other factors (RUDICH et al., 1979; WILLIAMS & SISTRUNK, 1979; KANISZEWSKY & RUMPLE, 1983; KANISZEWSKY et al., 1987).

The indication for nitrogen fertilization is based on the organic matter content of the soil. However, this is only an indication, and the crop's yield potential, the timing of the application, the crop's sensitivity to lodging, the previous crop and the amount of remaining straw, and the crop's stage of development should also be taken into account. In addition to climatic factors, such as the average temperature of the region and water availability (FLOSS, 2011).

Despite the various factors that influence the quantification of the dose of N to be applied, Souza & Moreira (2010) report that it is highly necessary to know the nutritional needs of the crop in order to obtain high yields and fruit quality, as tomato production is costly, mainly due to the need for high doses of fertilizers.

3.2.3 Nitrogen deficiency in tomato cultivation

Failure to supply tomato plants with adequate N can lead to nutritional deficiency. Nitrogen is easily redistributed in the plant via the phloem and, consequently, N-deficient plants first show symptoms in the old leaves, where there is a decrease in chlorophyll content. The longevity of the leaves can be altered by a lack of N, which is a mobile element and moves to the new parts of the plant, causing early senescence of the older parts (BUSATO, 2007).

The first symptom of nitrogen deficiency in tomatoes is chlorosis. This chlorosis is associated with reduced chlorophyll synthesis and changes in the shape of the chloroplasts (FLOSS, 2011). In addition to chlorosis in older leaves, nitrogen deficiency in tomato plants can also be expressed by the accumulation of anthocyanins (purpling) in stems, petioles and lower leaves (MARENCO & LOPES, 2005). This is due to the consequent accumulation of carbohydrates not used in N metabolism, which are used in pigment synthesis (TAIZ & ZEIGER, 2004).

Changes such as a reduction in the length, width and thickness of tomato leaves are also observed when there is an N deficiency in the plants (FURLANI, 2004). When we consider excess nitrogen in tomato plants, there is abundant leafing, with a dark green color and a high ratio between the dry mass of the aerial part and the dry mass of the roots (MARENCO & LOPES, 2005).

3.3 GROWING TOMATOES IN A PROTECTED ENVIRONMENT

3.3.1 Tomato production under protected cultivation

Cultivation in a protected environment is one in which one or more climatic factors are controlled, encompassing practices and technologies that provide safer and more protected plant management (WITTWER & CASTILHAS, 1995). When used correctly, the protected environment can provide yields two to three times higher than those obtained with field cultivation (CERMENO, 1990).

The growing demand for vegetables, especially tomatoes, with high quality throughout the year, including during the off-season, is contributing to the investment in new crop technologies. The use of plastic for the production of horticultural crops has been used in Brazil since the 1970s, when it began to be used for the production of strawberries (GOTO, 1997). The protected cultivation of crops has several benefits when compared to the conventional, open-air cultivation system, such as early production and the ability to grow crops in the off-season, higher productivity, greater use of nutrients, cleaner products, greater efficiency in the use of water and possible decreases in the incidence of diseases are some of them (FONTES, 1999).

Protected cultivation has been gaining ground among vegetable growers due to the greater ease with which growing conditions can be managed compared to the open field system (CARRIJO et al., 2004). Vegetable houses also allow the microclimate of the environment to be altered, making it possible to grow vegetables at times of the year that are not favorable for their development (MARTINS et al., 1994). They also provide greater productivity and better fruit quality (LOURES et al., 1998).

Evaluating the performance of two tomato hybrids (Sunny and EF-50), grown in a protected environment and in the field, they observed that the number of fruits was significantly influenced by the growing environment. The Sunny hybrid produced 31.2 fruits/plant in a protected environment and 15.0 fruits/plant in the field, while the EF-50 hybrid produced 26.8 and 12.5 fruits/plant, respectively (FONTES et al., 1997).

With the aim of evaluating the production of two tomato hybrids under four percentages of shading, Otoni et al. (2012) concluded that the shaded environments conditioned better agronomic performance for the tutored tomato plants, especially the 50% screen which provided higher total productivity (40.45 t ha^{-1}) compared to the open sky (12.90 t ha).$^{-1}$

According to the Brazilian Committee for the Development and Application of Plastics in Agriculture (Coblapa), in 2011 Protected Environment Production in Brazil was estimated at around 26,000 hectares. The state of Sao Paulo occupies more than 50% of the national area of this crop. In this context, there are three major agricultural production chains that use plastic for their cultivation: floriculture, fruit-growing and olericulture, the latter being the fastest growing (FIGUEIREDO, 2011).

3.3.2 Climatic factors

For tomatoes to thrive, the ideal conditions for growth and development must be taken into account. Dry, mild climatic conditions and good humidity are considered ideal, but extreme temperatures can interfere with plant hormones and the formation of flowers and pollen grains, as well as having an effect on fruit setting, coloring and ripening (LOPES & STRIPARI, 1998). According to Filgueira (2000), the tomato is a plant that requires temperatures with a difference of around 6 to 8°C between day and night, with temperatures between 21 and 28°C (day) and 15 to 20°C (night) being indicated as optimum. This variation also includes variations in the age of the plant and according to cultivar.

The tomato crop is extremely sensitive to weather conditions and when these are unfavorable in field conditions, they contribute to a considerable increase in its cultivation in protected environments (LOPES & STRIPARI, 1998). Protected cultivation has several advantages, one of which is the possibility of changing the meteorological elements induced by the greenhouse environment through its structure (plastic) (SOUZA & ESCOBEDO, 1997). Other advantages include increased productivity and fruit quality, production in the off-season, lower incidence of pests and diseases, greater uniformity of the final product and plant protection against frost and hail (RODRIGUES, 2015).

The use of plastic greenhouses modifies some climatic factors such as air temperature, relative humidity and solar radiation, thus influencing plant growth, development and production, compared to an open-air environment (BECKMANN et al., 2006). One of the elements that can be most influenced inside a greenhouse is the density of solar radiation, with a consequent reduction in evapotranspiration. Thus, protected cultivation has the advantage of reducing water consumption by the plants, when compared to the conventional cultivation system in the field (REIS et al., 2009).

Another factor that is highly influenced by the protected environment is the lower air exchange inside the greenhouse, thus reducing the exchange of sensible heat and water vapor between the internal and external environments (ATARASSI, 2004). Inside the greenhouse there is also a reduction in the ventilation rate, thus generating higher air temperatures during the day compared to the conditions in the field (PEZZOPANE et al., 1995).

In order to overcome the climatic limitations, Slater (1983) highlights growing in protected environments as a strategy, especially considering their efficiency in capturing radiant energy and the plants' better use of different climatic factors, thus providing significantly higher yields than in the field.

3.3.3 Shading in tomato growing

Although the tomato crop is a species that is widely capable of adapting to different weather conditions, there can be a reduction in its production and quality due to the occurrence of these adversities. In this sense, the use of black and aluminized shading screens (thermo-reflective) has been used to mitigate excess solar radiation and excess temperatures (SILVA et al., 2013).

In Juazeiro, BA, Rocha (2007) tested two tomato hybrids in five protected environments and under the open sky. In his study, he concluded that the use of shading screens reduced the intensity of radiant energy, improving the quality and distribution of solar radiation in protected environments. The crop performed better when compared to the open sky and the best performance of the hybrids was recorded in the diffuser screen and black shading screen.

Studying different levels of shading under the tomato crop, he found that 30% and 50% shading provided better production results, although there was no difference in terms of fruit quality. In his study, he also concluded that the use of shading reduces the intensity of radiation, providing better crop development (OTONI, 2010).

Comparing the differences in physical-morphological characteristics in the production of tomato seedlings under different shading screens, they concluded that the different shading screens had an effect on the microclimate and growth of tomato plants (SILVA et al., 2013). The 50% screen had the lowest photosynthetic flux density and the lowest air temperature, resulting in the best development of the aerial part of the tomato.

3.4 NITROGEN FERTILIZATION X SHADING INTERACTION

The tomato crop is one of the most demanding vegetables in terms of fertilization, and it has different demands depending on its stage of development, crop cycle, genotype and time of year. In protected cultivation, due to possible fluctuations in climatic factors and a reduction in the plants' transpiration rates, some nutrients may be less or more in demand by the crop (SILVA & GIORDANO, 200).

One of the first studies carried out to understand the absorption of nutrients by the tomato crop in a protected environment concluded that the nutrients absorbed most by the crop were potassium and nitrogen, followed by calcium, sulphur, phosphorus and magnesium. The authors also found that the nutrients nitrogen, potassium, sulphur and magnesium reached their maximum values between 100 and 120 days after germination, while calcium and phosphorus throughout the crop cycle (GARGANTINI & BLANCO, 1963).

Studying the absorption of nutrients by tomato plants under field and protected environment conditions, Fayad et al. (2002) concluded that, in a protected environment, absorption followed the growth of the plant and, during the fruiting period, the amount absorbed of all the nutrients increased. However, the quantities of nutrients required decreased from N, K, Ca, S, P, Mg, Cu, Mn, Fe to Zn.

In order to determine the dose of nitrogen to be applied to tomato plants in a protected environment, it was concluded that, in general, the efficiency of nitrogen use for agronomic purposes decreased as the amount of N applied increased, resulting in an excess of N in the soil (ARAÙJO et al., 2007). According to JOHSON & RAUN (2003), when there is excess N in the soil, there is a decrease in use efficiency, resulting in a loss of N from the soil-plant system.

The rational use of fertilizers not only reduces costs, but also guarantees production quality and minimizes environmental contamination (GOTO & TIVELLI, 1998). In this way, production in a protected environment with a shading screen can make better use of production resources, such as nitrogen fertilization, resulting in a reduction in the use of inputs.

4 MATERIAL AND METHODS

This chapter will describe the materials, methods and techniques used during the period of the experiment, characterizing the data needed to understand it, information on the location of the experiment, the experimental area, planning, implementation and conduct of the experiment, agronomic management, data collection, analysis and interpretation.

4.1 EXPERIMENTAL AREA

The experiment was carried out in the 2016 agricultural year in the greenhouse of the experimental area of the Biomonte Research and Development Company, in the district of Boca do Monte, in Santa Maria, Rio Grande do Sul, Brazil, whose geographical coordinates are 29°39.059' South and 53°57.413' West and altitude of 180.0 meters (Figure 2).

The experiment was conducted between the sowing of the seedlings on May 6, 2016 and the end of the fruit harvest and quality evaluations on November 15, 2016.

Figure 2 - Experiment site.

The region's predominant climate, according to the Koppen scale (MORENO, 1961), is characterized as humid subtropical (Cfa) with average temperatures of 18.9°C during the year. The average rainfall during the year is around 1700 mm.

4.1.2 Soil characteristics

The soil used for the experiment is classified as Argissolo Vermelho Distròfico Arnico. These are soils with a base saturation of < 50% in most of the first 100 cm of the AB horizon. According to MAPA Agricultural Zoning, the soil is considered type 2. SBCS textural class: Sandy loam, with 60.6% sand, 22.9% silt and 16.6% clay, according to the soil physics analysis report from the Soil Physics Laboratory - UFSM.

To determine the chemical characteristics of the soil, samples were taken from the 0-10 cm depth. The samples were sent to the BASELAB Soil Chemistry Analysis Laboratory for further interpretation and fertilizer recommendations for tomato cultivation. The chemical characteristics of the soil used in the experiment are shown in Table 1.

Table 1 - Chemical characteristics of Argissolo Vermelho Distròfico Arnico soil, Sao Pedro mapping unit - Average values from two replications. Santa Maria - RS, 2017.

Prof.(cm)	PH	M.O.		Exchangeable content g/100 g soil						
	H2O (1:1)	%	CTC	k molc.dm-3	Ca cmolc.dm-3	Mg cmolc.dm-3	Al	P mg.dm- 3	H + Al	SMP
0-10	4,885	1,7	6	0,328	1	0,8	1,32	17,2	4	6,08

4.2 EXPERIMENTAL DESIGN AND TREATMENTS

The experimental design used was the 2x5 bifactorial Completely Randomized Design (DIC) (Table 2), with five treatments in each environment (with and without a shading screen). In addition to the environment factor (factor A), the treatments were divided according to the dose of nitrogen fertilization (factor D) (equivalent to 0, 75, 150, 225 and 300 kg ha^{-1} of N), estimated according to the needs of the tomato crop and the analysis of the soil used in the experiment (Table 2). Each treatment consisted of twenty experimental units (pots), totaling 200 experimental units in the experiment.

Table 2 - Description of the treatments (factor A - shading and factor D - nitrogen doses) evaluated in the experiment. Santa Maria - RS, 2017.

Treatments	**Description**

A1D1	With shading screen and 0 kg ha^{-1} N With shading
A1D2	screen and 75 kg ha^{-1} N
A1D3	With a shading screen and 150 kg ha^{-1} N
A1D4	With a shading screen and 225 kg ha^{-1} N
A1D5	With a shading screen and 300 kg ha^{-1} N
A2D1	No shading screen and 0 kg ha^{-1} N
A2D2	No shade screen and 75 kg ha^{-1} N
A2D3	Without shading screen and 150 kg ha^{-1} N
A2D4	Without shading screen and 225 kg ha^{-1} N
A2D5	Without shading screen and 300 kg ha^{-1} N

The doses of nitrogen were divided into three periods: i) when the seedlings were transplanted into the pots; ii) 25 days after transplanting (first visible apical tertiary branch); and iii) 50 days after transplanting (full bloom), applied in solid form (urea - CO(NH $)_{22}$), in a half-moon around the tomato seedling, with equal doses for each period (Figure 3), according to Faria et al. (1996).

Figura 3 - Application of nitrogen doses divided into three periods. Santa Maria - RS, 2017. Photo: Morais, T. B. (2017).

Figura 4 The experimental area (greenhouse) used for the experiment contained 600 m^2 , measuring 50 x 12 m. The covering material for the greenhouse was high-density polyethylene with transparent activation, and this environment was considered to be the unshaded environment in this study. The shaded environment consisted, in addition to the polyethylene cover, of a black Sombrite® type shading screen with 50% coverage. Each environment (with and without shading) consisted of four rows of pots (experimental units), spaced 0.50 meters apart, randomly distributed and duly labeled according to their treatment (Figure 4).

Figura 5 - Experimental area (greenhouse). Santa Maria- RS, 2017. Photo: Morais, T.B. (2017).

4.3 SETTING UP AND CONDUCTING THE EXPERIMENT

To produce tomato seedlings, the seeds were sown in 128-cell expanded polystyrene trays containing commercial BIOPLANT® substrate. The trays were kept in a greenhouse and received adequate irrigation and phytosanitary treatment (Figure 5 A).

The Santa Cruz Kada (Paulista) cultivar was used. This cultivar has an indeterminate growth habit, a cycle of around 120 days, Santa Cruz type fruit, an average mass of 90 grams per fruit, a commercial length of 07 cm and a commercial diameter of 05 cm, suitable for industry and salad.

The seedlings were transplanted when they were on average 10 cm tall, 40 days after being sown in the trays. The seedlings were transplanted into black plastic pots, suitable for seedlings, with a capacity of 4.9 liters. 3,400 g of soil was added to each pot. To fill the pots, the soil was sieved, homogenized and corrected for acidity according to the soil analysis (Figure 5 B).

Figure 5 - Transplanting tomato seedlings into pots. Santa Maria - RS, 2017. Photo: Morais, T.B. (2017).

The pots were spaced as follows between rows: 0.30 m - 1.0 m - 0.30 m in the same room, to make it easier to assess the plants. The system used was vertical ties, one per plant, in a vertical direction, at a height of 2.00 meters. The plants were trained with a single stem and cut once a week.

During the seedling stage, the trays were placed in a mulching system with black plastic. After the seedlings were transplanted into the pots, irrigation was carried out as recommended by Pereira (2002), where the irrigation water was replaced in full (100% of the water consumed by the crop). Control was carried out using the weighing method. Initially, the pots were weighed at field capacity (previously determined by saturating the soil in the pots with water and, after three days, when they reached a constant weight, the weight corresponding to field capacity was taken). Before each irrigation, at least three pots corresponding to each treatment and each environment (30 experimental units) were weighed and the difference between the current weight and that corresponding to field capacity corresponded to the weight (volume) of the replacement water. The pots were weighed and the water replaced every 3 days.

Fertilization of the crop, excluding nitrogen fertilization, was in accordance with the needs and technical recommendations for tomato growing. The sources used were triple superphosphate and potassium chloride, 900 kg ha^{-1} of P and 215 kg ha^{-1} of K, via the soil. To complement the fertilization, calcium (10% Ca) and boron (0.5% B) were applied weekly via foliar application in the amount of 200 ml 100 L $.^{-1}$

For phytosanitary control, sprays were made with specific products registered for tomato growing in the state of Rio Grande do Sul. The following active ingredients were used for whitefly control: beta-cyfluthrin 12.5 g a.i. + imidacloprid 100 g a.i., acetamiprid 200 g a.i., applied when the level of economic damage to the pest was reached. For pathogen control, the following active ingredients were used: azoxystrobin 500 g a.i., chlorothalonil 500 g a.i., metiram 550 g a.i. + pyraclostrobin 50 g a.i., azoxystrobin 200 g a.i. + difenoconazole 125 g a.i., applied preventively, from the moment the seedlings were transplanted into the pots, to avoid the presence of pathogens during the experiment. In this study, the applications took place at an interval of 14 days. No herbicide applications were necessary during the experiment to control weeds.

4.4 VARIABLES ANALYZED

4.4.1 Climate variables

During the course of the experiment, the temperature (T°C) and relative humidity (RH%) values were checked daily in the two environments (with and without shading), and measured using a Digital Thermo-Hygrometer with Extension Cable, model 1566-1.

In addition to temperature (T°C) and humidity (RH%), daily light assessments were carried out in the two environments (shaded and unshaded). The values were determined using a Lutron LX-101 digital lux meter. The lux meter was positioned on a horizontal plane in both environments, where it was read in lux.

4.4.2 Plant height

Plant height was determined using a graduated ruler, with values expressed in centimeters (cm). The measurements were taken weekly, starting the week after the seedlings were transplanted into the pots and ending 10ª weeks after transplanting (70 DAT). Its value corresponded to the distance between the base of the plant, starting from the ground, and the top of the plant (Figure 6), according to Otoni et al. (2012).

4.4.3 Plant diameter

The stem diameter values of the tomato plants were obtained, expressed in cm, using a professional 150 mm manual caliper. The values were determined in the basal region of the plant, close to the ground (Figure 6), according to Campos (2013).

4.4.4 Leaf area

Leaf area was determined by measuring the length (C) and width (L) of the leaves of five marked plants per treatment. According to Reis et al. (2013), length is defined as the distance between the point where the petiole inserts into the leaf limb and the opposite end of the plant. Width was defined as the largest dimension perpendicular to the length axis (Figure 6) and values were expressed in cm .[2]

Figure 6 - Evaluation of plant height, plant stalk diameter and leaf area. Santa Maria - RS, 2017. Photo: Morais, T.B (2017).

They were evaluated weekly, starting in the third week after transplanting (21 DAT) until the eighth week (56 DAT), when the values stabilized. The procedure adopted by Ashkey et al. (1963) was used to calculate leaf area:

AF = CLf

AF = leaf area, in cm^2

C = leaf length in cm

L = leaf width, in cm f = correction factor (estimated for tomatoes as 0.59)

4.4.5 Chlorophyll content

For the indirect chlorophyll measurements, the method was carried out on the living plant without damaging it (non-destructive method), using the FALKER CFL 1030 chlorophyll device, which indirectly determines the chlorophyll concentration of the leaves by the reflectance of green at a wavelength of approximately 650 nm (MONTEIRO, 1999). The evaluation was carried out on the leaves most exposed to sunlight, fully expanded, without signs of senescence and healthy, at 30, 50, 70 and 90 DAT. The results for chlorophyll *a b* and total chlorophyll were expressed as ICF (Falker Chlorophyll Index) (Figure 7).

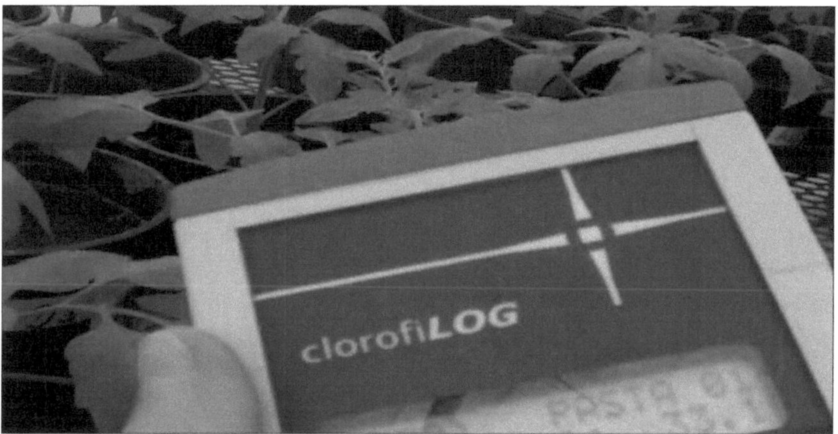

Figure 7 - Chlorophyll assessment using the non-destructive method. Santa Maria - RS, 2017. Photo: Morais, T. B. (2017).

4.4.6 Stomatal behavior

Semi-permanent slides were made to assess the leaf epidermis from the front. The epidermal printing technique was used, where a drop of universal instant adhesive (cyanoacrylate ester) is placed on a glass slide. The region of interest of the leaflet is then pressed onto the slide for approximately 10 seconds, thus allowing the leaflet to separate from the slide and maintaining the impression of the epidermis (SEGATTO et al., 2004). The evaluation consisted of 5 random plants per treatment, where two slides were made per plant (abaxial and adaxial side of the leaflet), always evaluating the leaflet with the same location on the plants, at 2pm (Figure 8).

An optical microscope with a 15x eyepiece and 40x objective was used to observe the slides. The images used were captured using a microscope capture system equipped with a digital camera, where the size of the captured image (real visible field) was 0.272 mm^2 . Stomata were counted and their diameters assessed using the Axio Vision® driver. Stomatal density (number of stomata/mm^2), polar diameter (PD) and equatorial diameter (ED) expressed in μm and stomatal functionality (PD/ED ratio) were evaluated.

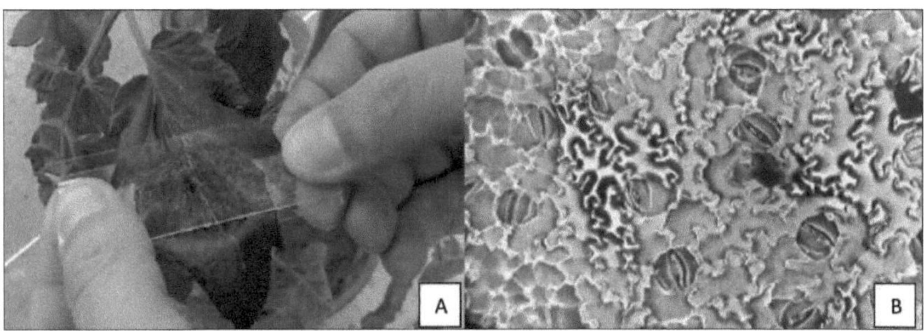

Figure 8 - Making the semi-permanent slides (A). Front view of the abaxial side (600x magnification) of a tomato leaflet (B). Santa Maria - RS, 2017. Photo: Morais, T.B. (2017).

4.4.7 Aerial part dry mass

Based on the samples collected from the aerial part of the treatments, the leaf dry mass was evaluated. The samples were dried in an oven at a constant temperature of 70°C until they reached a constant weight. They were then weighed on a 2200 g (0.01 g) EDULTEC digital precision balance and expressed in grams (g).

4.4.8 Total fruit production

The fruit was harvested by hand and placed in plastic bags labeled with the respective treatment. After harvesting, the fruit was weighed on a 2200 g (0.01 g) EDULTEC digital precision scale, and the length and height of the fruit, expressed in cm, were measured using a 150 mm manual caliper (Figure 9). After the evaluations, the fruit was kept refrigerated so that there was no loss of fruit quality attributes.

Harvesting began 17ª weeks after the seedlings were transplanted into the pots. Fruits with stage 5 characteristics (light red), according to the scale proposed by PAULA J.T. (2012), were harvested once a week for three weeks after the start of the harvest.

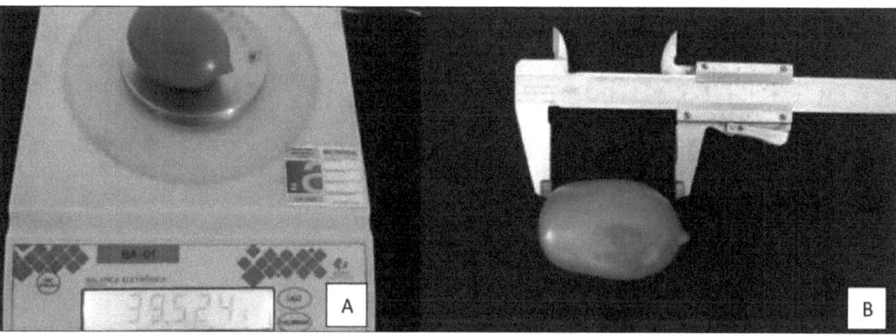

Figure 9 - Assessment of fruit weight on a digital scale (A). Evaluation of fruit length and height using a 150 mm caliper (B). Santa Maria - RS, 2017. Photo: Morais, T.B. (2017).

4.4.9 Fruit quality

For the fruit quality assessments (pH, soluble solids (SS) and titratable acidity (TA)), samples of tomato fruit pulp (around 200 grams) were prepared. The samples were prepared using a Mondial Power 2 Black blender (Figure 10 A and B). The fruits were chosen at random in order to form a homogeneous sample of each treatment.

4.4.9.1 pH

The pH of the tomato pulp was assessed using a portable digital pH meter with a 0.0-14 pH scale, 0.1 resolution, accuracy of $\leq \pm$ 0.03 pH and 2-point calibration (Figure 10 C) (OTONI et al., 2012).

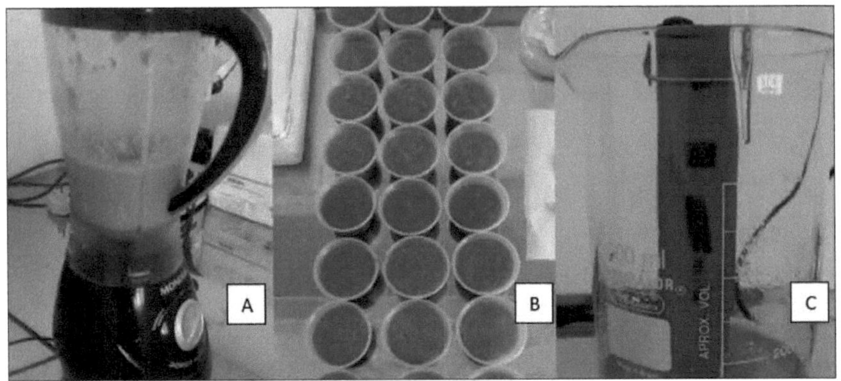

Figure 10 - Sample preparation of tomato fruit pulp juice (A). Samples separated by treatment (B). Evaluation of juice pH (C). Santa Maria - RS, 2017. Photo: Morais, T.B. (2017).

4.4.9.2 Soluble solids (SS)

The soluble solids content was measured using an Abbe Type WYA Model 2WA-J table refractometer and the result was expressed in degrees Brix (Figure 11 A) (MORETTI, 2006).

4.4.9.3 Titratable acidity (TA)

Acidity was determined by titrating 10 ml of tomato juice with 0.1 N NaOH until it reached pH 8.2, using a graduated manual titrator (burette) and a TE 0851 tabletop magnetic stirrer. Titratable acidity was expressed as a percentage, assuming citric acid as the predominant acid in tomato juice, according to IAL (2008) (Figure 11 B and C). Total titratable acidity (ATT) was expressed in grams of acid per 100 grams or 100 mL of sample, using the formula:

$$ATT_{(g/100\ ml)} = \frac{n \text{ X } N \text{ X } Eq}{10 \text{ X } V}$$

N = normality of sodium hydroxide solution

n = volume of sodium hydroxide solution used in the titration in ml p = mass of sample in grams

V = sample volume in ml

Eq = gram equivalent of acid

4.4.9.4 SS/AT

Soluble solids/titratable acidity ratio (SS/AT) - calculated using the Soluble Solids/Titratable Acidity ratio, according to IAL (2008).

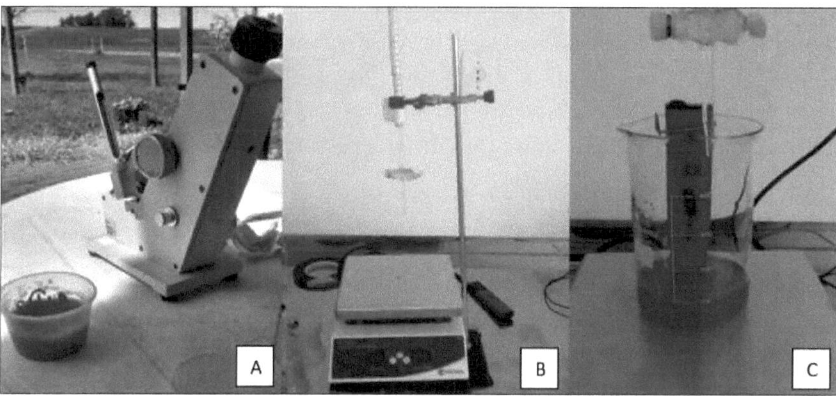

Figure 11 - Evaluation of Soluble Solids (°Brix) in tomato pulp juice (A). Titratable acidity assessment (B and C). Santa Maria - RS, 2017. Photo: Morais, T.B. (2017).

4.5 STATISTICAL ANALYSIS

Statistical analysis was carried out comparing each environment in isolation (with and without a shading screen) and for each sampling period. The experiment is considered to be qualitative-quantitative, so regression was carried out for the quantitative variables and analysis of the separation of means for the qualitative variables. The statistical analyses were processed using the SISVAR 5.0 statistical analysis program and the graphs were generated using the SigmaPlot program.

5 RESULTS AND DISCUSSION

5.1 CLIMATE ASSESSMENTS

Throughout the experiment, there was a variation in minimum and maximum temperatures between the two environments under study (shaded and unshaded). The average daily temperature in the unshaded environment was higher than in the shaded environment, with averages of 20.9°C and 20.5°C, respectively. These values are within the ideal range for the crop which, according to Filgueira (2008) during the vegetative growth, flowering and fruit ripening phases, is between 18 and 24°C (Figure 12).

The relative humidity (RH%) showed no difference between the two environments (with and without shading), with an average of 75.71% over the course of the experiment (Figure 3). In this sense, according to Guimaraes et al. (2007), the most suitable daily relative humidity range for this crop is between 50 and 75% (Figure 12).

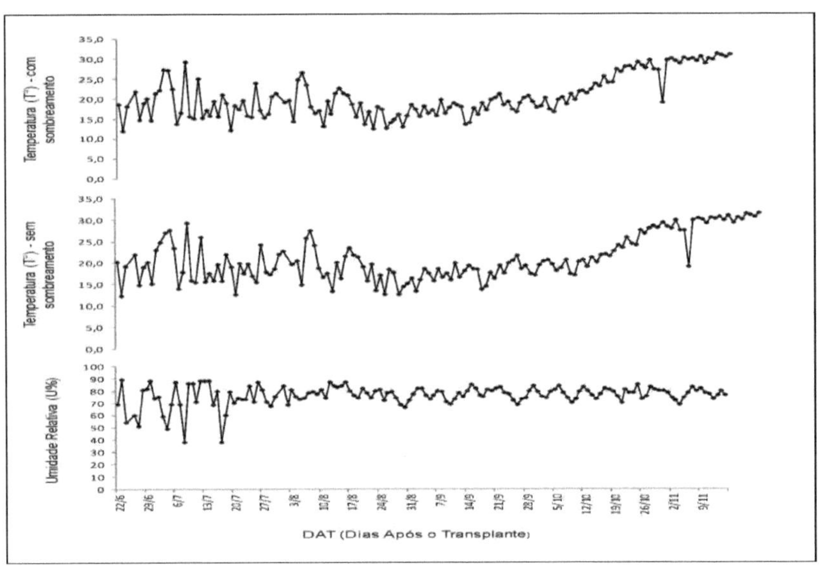

Figure 12 - Average daily temperatures in the rooms (with and without shading) and relative humidity. Santa Maria - RS, 2017.

The environment without 50% black screen shading had an average value of 4082.60

lumens, while the environment with shading had an average of 1678.69 lumens (Figure 13). The higher the light value (lumen), the more efficient the environment for growing tomatoes, because, according to Papadopoulos et al. (1997), Andriolo (2000), light is essential for the first stage of the CO_2 fixation chain, photosynthesis, where the biochemical energy necessary for the growth and production of the crop is produced.

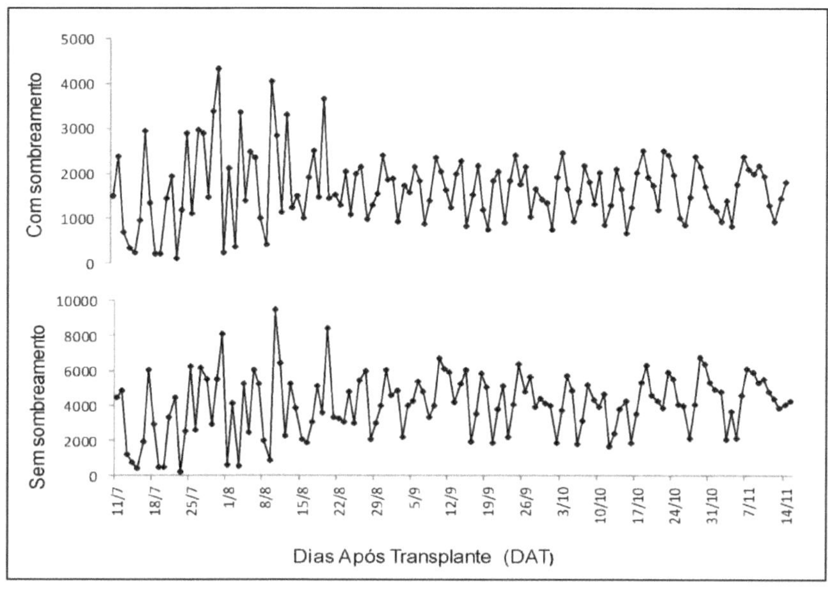

Figure 13 - Luminosity measurements in the two environments, measured in lumens. Santa Maria - RS, 2017.

5.2 PLANT HEIGHT AND DIAMETER

According to the results of the analysis of variance, shown in Figure 14, there was a significant effect ($p \leq 0.05$) of the treatments on the plant height variable in the unshaded environment in all the evaluations carried out. However, in the shaded environment, there was no significant effect at 7 DAT (days after the treatments were applied), but in the other evaluations (14, 21, 28, 35, 42 and 49 DAT) there was a significant effect ($p \leq 0.05$).

The regression analysis for plant height (Figure 14) showed that the data fitted a quadratic function. There was an increase in plant height as a function of the doses of nitrogen applied when compared to the treatment without application, in both environments (with and without shading) (Figure 14). Similar results were reported by Mehmood et al. (2012) where they observed this same increase according to the increase in the dose of N applied and the treatment without nitrogen application provided the lowest plant height. Porto (2013) in his study on the source and dose of nitrogen in the production and quality of tomatoes concluded that when the crop was subjected to increasing doses of N, it showed an increase in plant height.

Plant growth as a function of nitrogen dose may be related to the availability of the element to plants, given its importance for their development. Nitrogen is a constituent of many plant compounds, including all proteins (formed from amino acids) and nucleic acids, relating to the formation of DNA and RNA, regulation of the photosynthetic rate, resulting in increased cell division and plant growth (HAQUE et al., 2011).

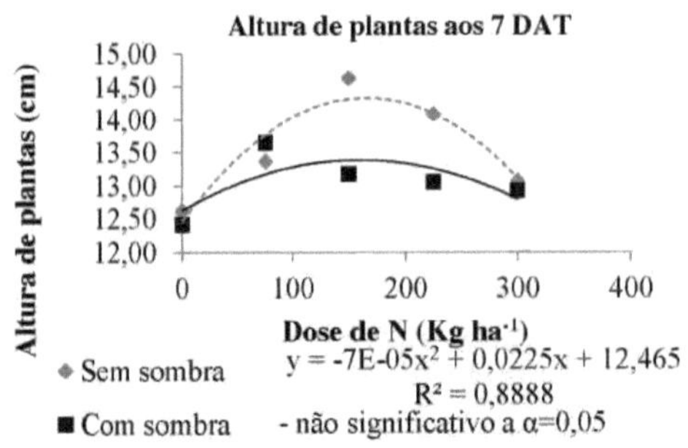

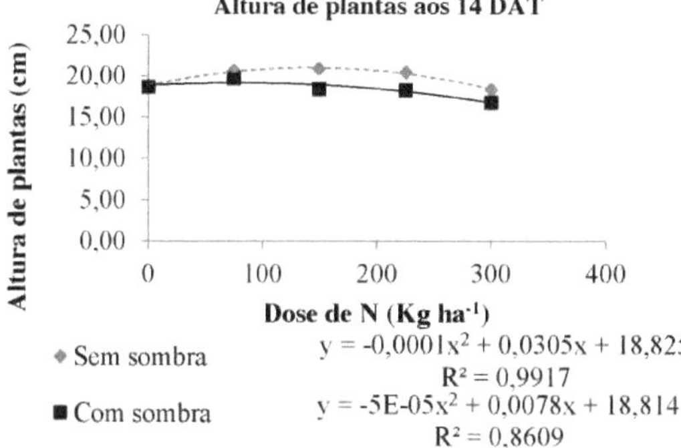
Altura de plantas aos 14 DAT
Altura de plantas (cm)
25,00
20,00
15,00
10,00
5,00
0,00
0
100
200
300
400
Dose de N (Kg ha-1)
Sem sombra
y = -0,0001x2 + 0,0305x + 18,825
R² = 0,9917
Com sombra
y = -5E-05x2 + 0,0078x + 18,814
R² = 0,8609

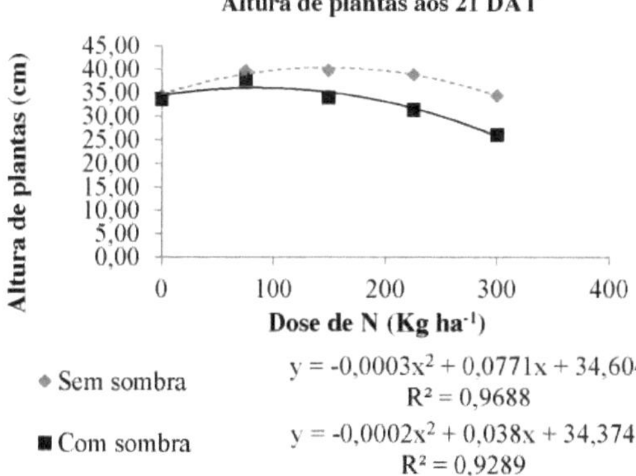
Altura de plantas aos 21 DAT
Altura de plantas (cm)
45,00
40,00
35,00
30,00
25,00
20,00
15,00
10,00
5,00
0,00
0
100
200
300
400
Dose de N (Kg ha-1)
Sem sombra
y = -0,0003x2 + 0,0771x + 34,604
R² = 0,9688
Com sombra
y = -0,0002x2 + 0,038x + 34,374
R² = 0,9289

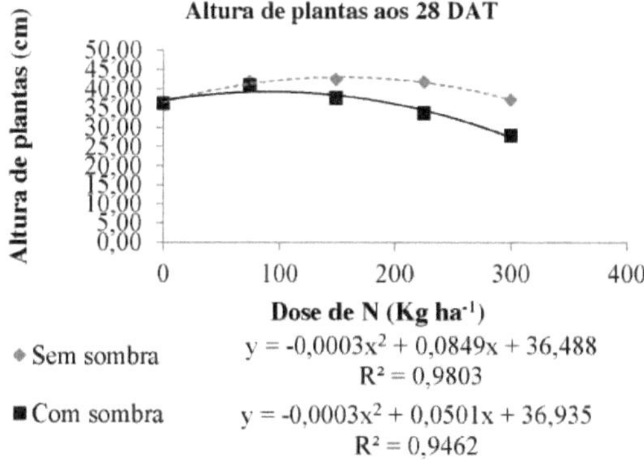
Altura de plantas aos 28 DAT
Altura de plantas (cm)
50,00
45,00
40,00
35,00
30,00
25,00
20,00
15,00
10,00
5,00
0,00
0
100
200
300
400
Dose de N (Kg ha-1)
Sem sombra
y = -0,0003x2 + 0,0849x + 36,488
R² = 0,9803
Com sombra
y = -0,0003x2 + 0,0501x + 36,935
R² = 0,9462

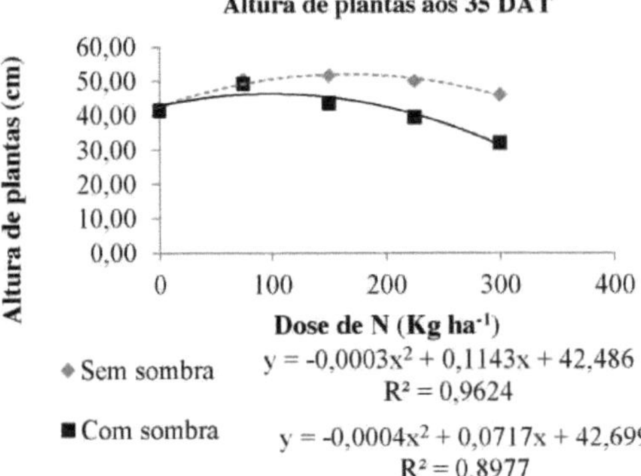

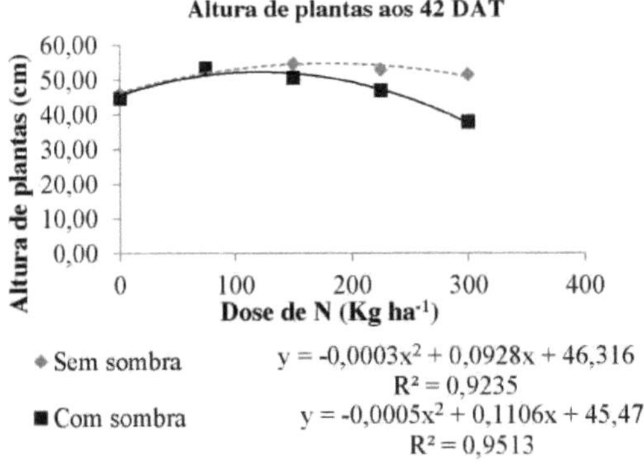

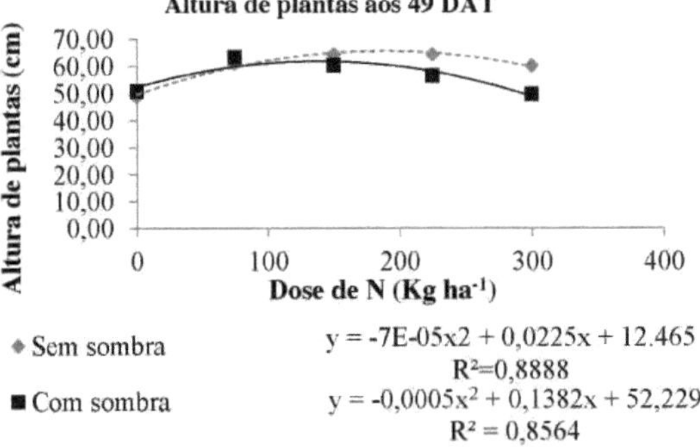

Figura 14 - Evaluation of plant height (cm) when subjected to increasing doses of nitrogen in two growing environments (with shading and without shading). Santa Maria, RS, 2017.

As for the analysis of plant stem diameter, the results found for the influence of nitrogen doses in relation to environments with and without shading on the tomato crop for all the DAT evaluated were significant ($p \leq 0.05$) (Figure 15).

The results for plant stem diameter also fitted a quadratic function (Figure 15). In all the evaluations carried out, there was an increase in plant stem diameter according to the dose of N applied, when compared to the treatment without N application, regardless of the environment in which it was subjected. These results are in line with those obtained by Porto (2013), who found lower growth in the treatment without the N dose. Nitrogen generally promotes an increase in plant vigor (ADAMS, 1986; PAPADOPOULOS, 1991), which is associated with plant height and stem diameter (NAVARRETE et al., 1997).

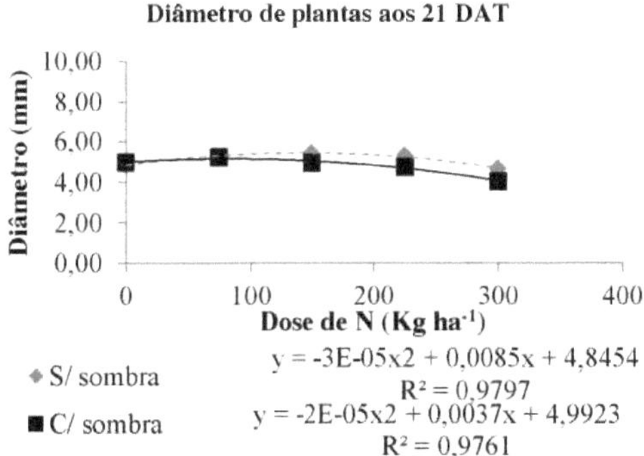

Diâmetro das plantas aos 28 DAT

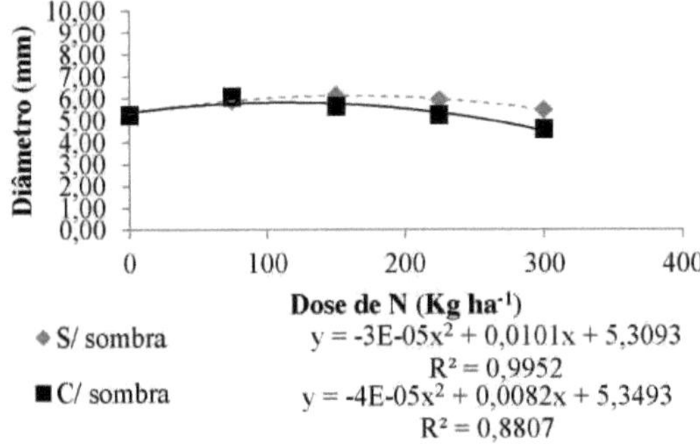

Diâmetro das plantas aos 35 DAT

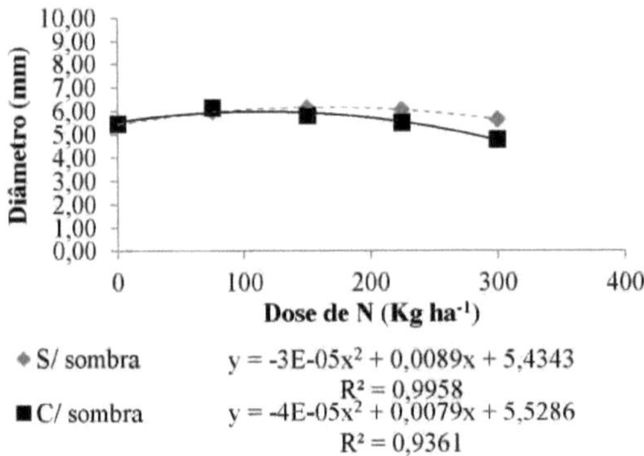

Diâmetro das plantas aos 42 DAT

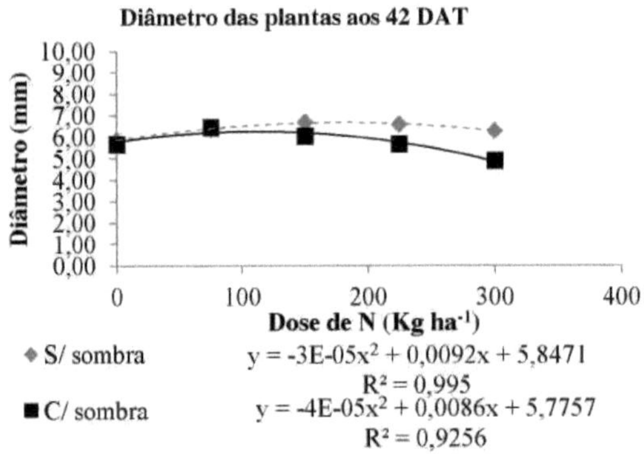

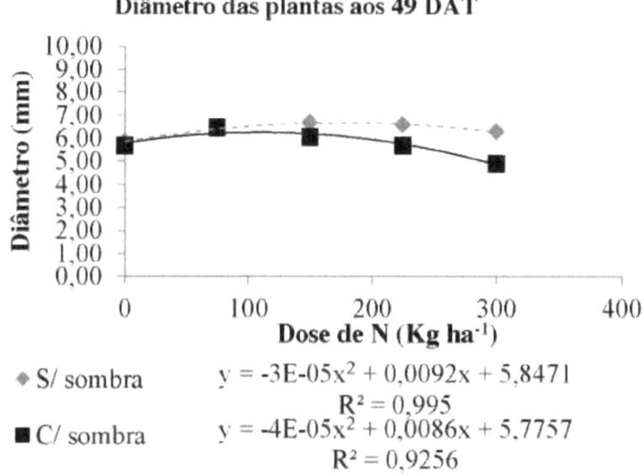

Figura 15 - Evaluation of plant stem diameter (cm) when subjected to increasing doses of nitrogen in two growing environments (with shading and without shading). Santa Maria, RS, 2017.

Analyzing Figures 14 and 15, maximum technical efficiency (MET) was observed with the dose of 154 kg ha^{-1} of N in the unshaded environment and MET of 110 kg ha^{-1} of N in the shaded environment, in the evaluation carried out at 42 DAT (days after transplanting), in relation to plant height and MET of 184 kg hà-1 in the unshaded environment and 172 kg hà-1 in the shaded environment, in relation to plant diameter. The doses subsequent to these showed a decrease in the values of the variables analyzed in the ratio of 2.27% and 9.00% in the unshaded environment (225 and 300 kg ha^{-1} of N, respectively) and 6.79, 12.96 and 24.06% in the shaded environment (150, 225 and 300 kg ha^{-1} of N, respectively) (Figure 14 and 15). The reduction in these values when higher doses of N were applied can be explained by the excess of nitrogen in the plant, which can lead to nutrient toxicity and, in extreme cases, plant death.

Nitrogen is mainly absorbed in the form of $NO3^-$ and $NH4^+$, but the form in which it is absorbed can lead to a reduction in the absorption of other cations such as Ca and Mg, as in the case of NH_4^+ . In tomatoes, the initial absorption of N by the plant is in the form of NO3" and $NH4^+$, but the oxidized form of N must be reduced before it can enter the plant metabolism, a process that takes place in the leaves (CASTRO et al., 2005). An inadequate or excessive supply of N therefore has a negative and immediate influence on a plant's metabolism, and the same author points out that nitrogen compounds are involved in the long-distance transportation of micronutrients (Mn and Cu, for example), leading to an indirect deficiency of these elements in the plant.

According to Taiz & Zeiger (2009), the absorption of ammonium and nitrate ions at high concentrations in the soil after fertilization can exceed a plant's capacity to assimilate these ions, leading to their accumulation in plant tissues, which is considered toxic to plants.

Figures 14 and 15 show that the lowest dose of N applied (138.2 kg ha^{-1} of N) and subjected to the shaded environment reached the maximum value for tomato plant height, at 49 DAT, this value reaching 53.25 cm. This result may be related to the lower value of the light compensation point (balance between photosynthetic absorption and CO_2 release) with a consequent lower respiration rate and photosynthetic rate (TAIZ & ZEIGER, 2009). In this way, there was less loss in terms of utilization of the nitrogen applied, and the occurrence of toxicity in tomato plants due to excess N was seen with one dose of the nutrient applied.

In all the evaluations carried out, the unshaded environment showed the highest values for plant height and diameter, regardless of the dose of nitrogen applied. These results may be related to the need for light on nutrient assimilation, as in the case of nitrogen, since nitrate reduction requires high light availability. For this reason, when plants are subjected to long periods of low light intensity, it is possible to observe low assimilation of nutrients and possible symptoms of nutritional deficiency (FLOSS, 2011). Also according to Floss (2011), the low growth of shaded plants is due to the lower light intensity and lower light quality.

5.3 LEAF AREA

Figure 16 shows the evolution of leaf area throughout the crop cycle. Growth occurred up to 56 DAT. Fayad et al. (2001) found a similar behavior for the Santa Clara cultivar, showing stable leaf area up to 58 DAT and a subsequent decline in growth.

According to the results of the analysis of variance, shown in Figure 16, there was a significant effect ($p \leq 0.05$) of the treatments on the leaf area (LA) variable in both environments for all the DAT evaluated. The results obtained from the regression analysis relating to AF (Figure 16) show that only the result for the shaded environment in the evaluation at 28 DAT showed a linear function, the others fitted a quadratic function.

In the unshaded environment, the PA values increased with the application of nitrogen, when compared to the treatment without N. However, from the dose of 194.34 kg ha^{-1} of N (MET) onwards, there was a reduction in the values found at 56 DAT. This can be attributed to the occurrence of toxicity caused by $NH4^+$, reducing plant development, as well as the negative influence on the supply of cations for plant development, such as calcium and magnesium (BORGOGNONE et al., 2013). Borgognone et al. (2013) also observed that leaf area was reduced when $NH4^+$ was supplied as the exclusive source of nitrogen.

In the shaded environment, the treatment with 123.29 kg ha^{-1} of N provided the highest leaf area (LA) averages in the 56 DAT evaluation, after which subsequent doses of N application showed a reduction in values (Figure 16). This reduction is attributable to the same toxicity found in the unshaded environment, caused by the excess nitrogen applied. The occurrence of toxicity with the application of a lower dose of N, compared to the unshaded environment, is due to the lower photosynthetic and respiration rates of the plants in the shaded environment, resulting in lower losses in the use of the nutrient (TAIZ & ZEIGER, 2009).

With regard to the environments, it was found that in the environment with 50% shade, the treatments with no added nitrogen and 123.29 kg ha^{-1} of N caused greater leaf area when compared to the environment with no shade (Figure 16). These results may be linked to the crop's need to increase its photosynthetic area to compensate for the lack of light. According to Scalon et al. (2001), these results are in line with what is normally observed, since there is a need to increase the photosynthetic surface to maximize light absorption in shaded environments. The expansion of the leaves under low light conditions indicates a way for the plant to compensate for the reduced luminosity, causing greater utilization as the surface area increases (CAMPOS & UCHIDA, 2002). However, in the other nitrogen doses applied (150, 225 and 300 kg ha^{-1}), the environment without shading showed higher leaf area values. This is due to the reduction in plant development in the shaded environment, possibly due to toxicity from the excess nitrogen applied.

Área foliar aos 21 DAT

Área (mm²): 700,00; 600,00; 500,00; 400,00; 300,00; 200,00; 100,00; 0,00

0 100 200 300 400

Dose de Nitrogênio (Kg ha^{-1})

◆ Sem sombra $y = -0.0083x2 + 2.1989x + 482.3$ $R^2 = 0.8541$

■ Com sombra $y = -0.0076x2 + 1.5264x + 504.14$ $R^2 = 0.8671$

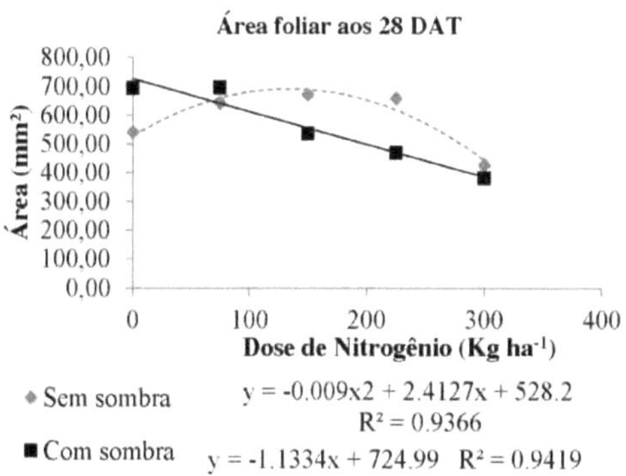

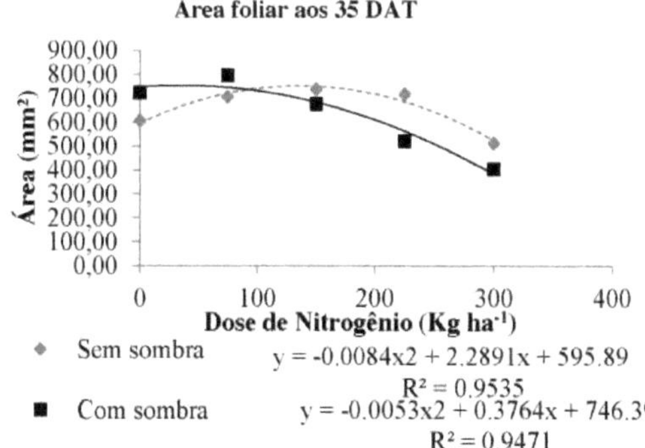

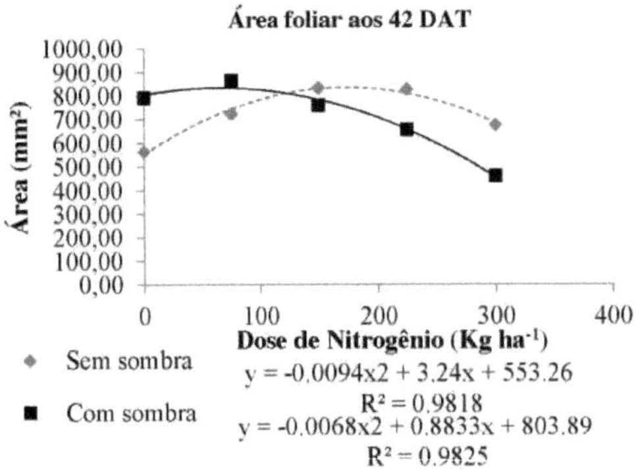

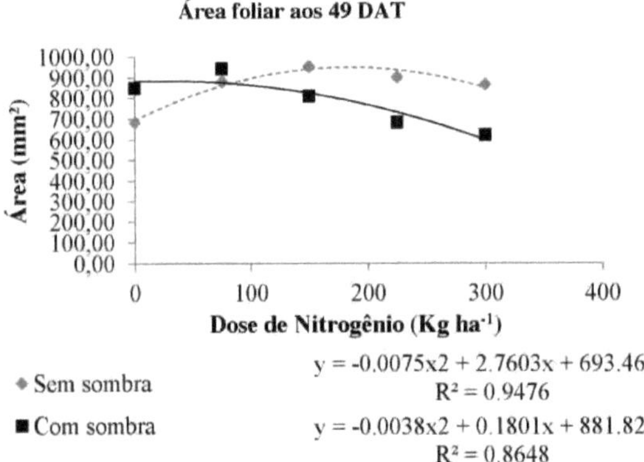

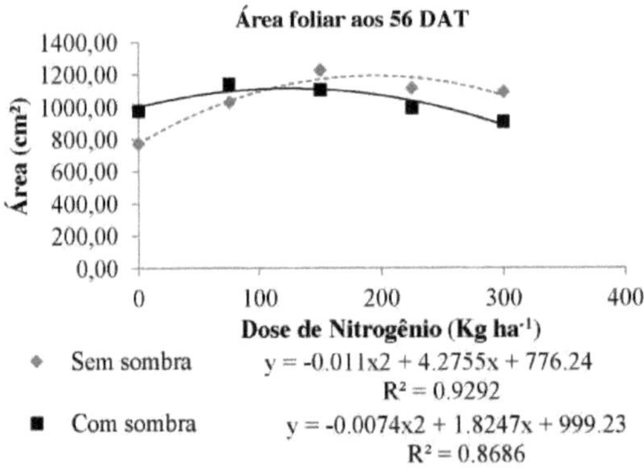

Figure 16 - Evaluation of leaf area (mm2) when subjected to increasing doses of nitrogen in two growing environments (with shading and without shading). Santa Maria, RS,

2017.

5.4 CHLOROPHYLL CONTENT

The chlorophyll *a b* and total (ICF) content were significantly affected ($p \leq 0.05$) by the doses of N and the environment in which the plants were located (Figure 17). They showed similar behavior, with quadratic responses in most of the evaluations as a result of the increase in N doses and in relation to the environment.

In both environments (shaded and unshaded) the total chlorophyll content increased as the dose of nitrogen applied increased, with the dose of 291 kg ha^{-1} of N (MET) showing the highest results in terms of chlorophyll content in the unshaded environment and 303 kg ha^{-1} in the shaded environment (MET). These results corroborate those obtained by other authors in pumpkin (SWIADER & MOORE, 2002), potato (GIL et al., 2002), tomato (FERREIRA et al., 2006) and other crops, where increases in total chlorophyll content were found as the dose of N applied increased. However, it was observed that in the maximum dose of N applied (300 kg ha^{-1}) in the evaluation at 30, 70 and 90 DAT, the increase in chlorophyll content was lower than in the previous dose, which was due to the toxicity caused by the excess of the nutrient, resulting in lower photosynthetic activity in the plant.

Nitrogen is a nutrient that participates in the synthesis and structure of chlorophyll molecules, so increasing the supply of N to plants, up to a certain limit, increases chlorophyll levels and the intensity of green color in plant leaves (FONTES & ARAUJO, 2007).

The maximum chlorophyll content values were found in the evaluation carried out at 60 DAT, with the ICF being 46.625 (unshaded environment with 228 kg $_{ha-1}$ of N applied) and 46.950 (shaded environment with 244.5 kg $_{ha-1}$ of N applied) when the tomato plants had all three doses of nitrogen applied, according to the treatments. After this evaluation, the chlorophyll content decreased. This decrease may be associated with the plants' physiology, in which they prioritize the development of reproductive organs, directing photoassimilates to reproductive structures from 60 DAT onwards. Maia (2011), working with bean cultivars and using the portable chlorophyll meter to determine nitrogen fertilization, found that in most of the cultivars studied, the maximum nitrogen levels were observed until the reproductive development stage, after which they began to decrease.

In relation to the environments, the shaded plants had higher levels of chlorophyll *a b* and total chlorophyll than the plants in the unshaded environment (Figure 17). These results corroborate those found by Junior et al. (2005) in relation to chlorophyll pigments, where they observed the highest values of total chlorophyll and chlorophyll *a* in *Cupania vernalis* leaves subjected to 50% and 70% shading. Coelho et al. (2013) analyzed the physiological responses of cowpea varieties subjected to different shading levels, concluding that 50% shading provided higher chlorophyll *a* indices when compared to full sun conditions. Souza et al. (2011) also observed higher levels of chlorophyll *a* and *b* when the guaco plants were grown under screens with greater light retention.

According to Taiz & Zeiger (2004), chlorophyll biosynthesis is a light-dependent biochemical reaction. However, at higher radiation intensities, chlorophyll molecules are susceptible to photoxidative processes, which are balanced at lower radiation levels. In this way, shaded leaves generally have a higher chlorophyll concentration than those grown in full sun (CASTRO, 2002; ALVARENGA et al., 2003). The reduction in chlorophyll content at higher levels of radiation (less shade) is widely reported in the literature, as recorded by ATROCH et al. (2001), KITAJIMA & HOGAN (2003) and ALVARENGA et al. (2003).

Figure 17 shows a small increase in chlorophyll *b* in the shaded environment at all nitrogen doses and DAT. According to Scalon et al. (2002), the increase in chlorophyll *b* in leaves subjected to low light levels is considered an important characteristic because chlorophyll *b* captures energy from other wavelengths and transfers it to chlorophyll *a*, which effectively acts in the photochemical reactions of photosynthesis and represents a mechanism for adapting to lower light intensity conditions.

Another factor that should be considered when assessing the chlorophyll content of plants is the ratio between chlorophyll *a* and chlorophyll *b*. The chlorophyll *a/b* ratio is directly related to the plants' ability to maximize light capture in conditions of greater shading (CRITCHLEY, 1999). According to Table 3, the shaded environment showed the highest chlorophyll *a/b* values from the evaluation carried out at 40 DAT onwards, regardless of the dose of N applied. Taiz & Zeiger (2009) state that shaded leaves have more chlorophyll per reaction center, the ratio between chlorophyll *b* and chlorophyll *a* is higher and they are generally thinner than sunny leaves.

Table 3 - Evaluation of the chlorophyll *a/b* ratio when subjected to increasing doses of nitrogen in two growing environments (with shading and without shading). Santa Maria, RS, 2017.

	Chlorophyll a/b ratio							
Treatments	Without[1] 20 DAT	With[2] 20 DAT	Without 40 DAT	At 40 DAT	Without 60 DAT	At 60 DAT	Without 80 DAT	With 80 DAT
0 kg ha^{-1} N	3,151a	3,212a	3,071a	3,263a	2,950a	2,974a	3,079a	3,072a
75 kg ha^{-1} N	3,420a	3,425a	3,143a	3,150a	2,488ab	2,808b	3,031b	3,069a
150 kg ha^{-1} N	3,399a	3.484ab	2,894ab	2,920ab	2,260abc	2,536c	2,494c	2,833b
225 kg ha^{-1} N	3,066a	3,231b	2,860ab	2,867b	2,147bc	2,202d	2,357d	2,469c
300 kg ha^{-1} N	3,512a	3,551b	2,634b	2,722b	2,029c	2,152c	2,295e	2,469c

[1] Treatments applied to tomato plants in an unshaded environment.

[2] Treatments applied to tomato plants in a shaded environment.

*Means not followed by the same letters in the columns differ by Tukey's test at 5% probability of error.

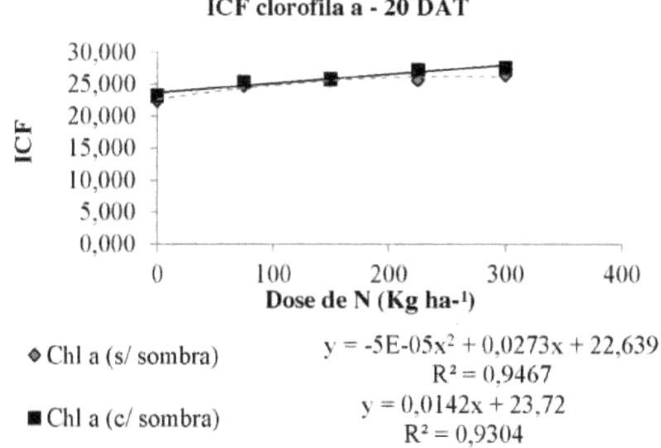

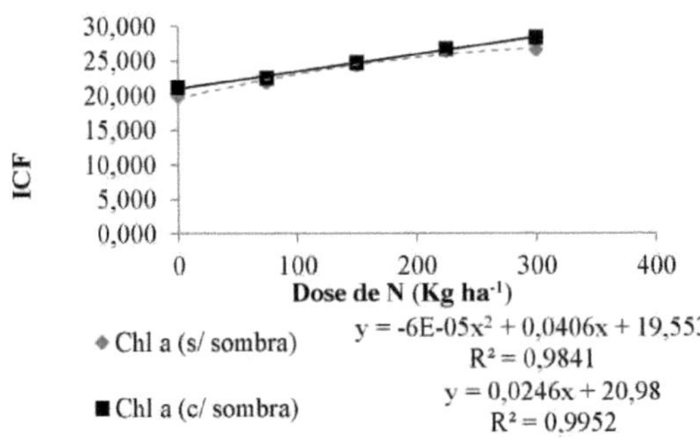
ICF clorofila a - 40 DAT
30,000
25,000
20,000
15,000
10,000
5,000
0,000
ICF
0
100
200
300
400
Dose de N (Kg ha-1)
Chl a (s/ sombra)
y = -6E-05x² + 0,0406x + 19,553
R² = 0,9841
Chl a (c/ sombra)
y = 0,0246x + 20,98
R² = 0,9952

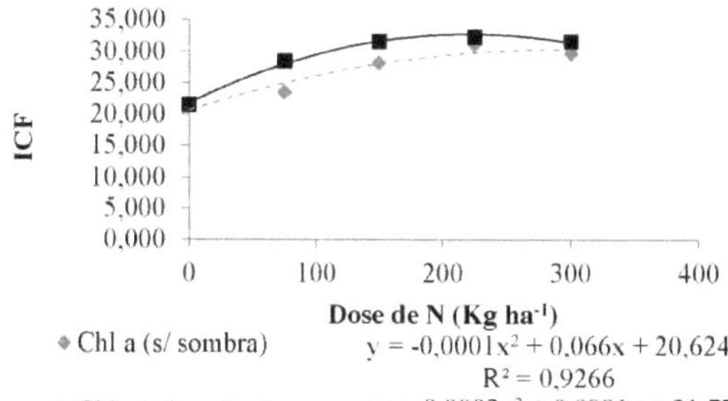
ICF clorofila a - 60 DAT
35,000
30,000
25,000
20,000
15,000
10,000
5,000
0,000
ICF
0
100
200
300
400
Dose de N (Kg ha-1)
Chl a (s/ sombra)
y = -0,0001x2 + 0,066x + 20,624
R² = 0,9266
Chl a (c/ sombra)
y = -0,0002x2 + 0,0991x + 21,774
R² = 0,9927

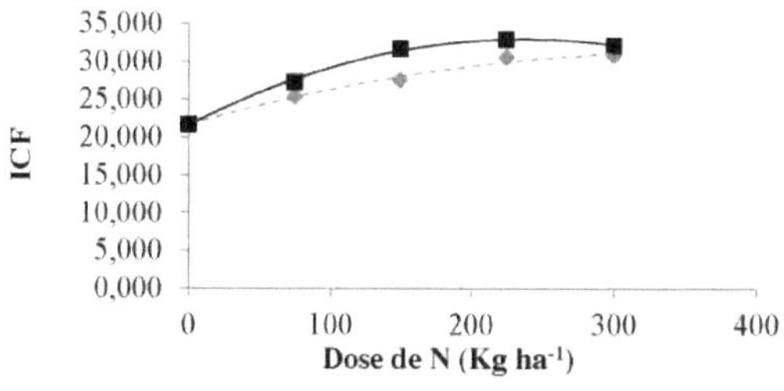
ICF clorofila a - 80 DAT
35,000
30,000
25,000
20,000
15,000
10,000
5,000
0,000
ICF
0
100
200
300
400
Dose de N (Kg ha-1)
Chl a (s/ sombra)
y = -7E-05x2 + 0,0534x + 21,669
R² = 0,9878
Chl a (c/ sombra)
y = -0,0002x2 + 0,0959x + 21,565
R² = 0,9977

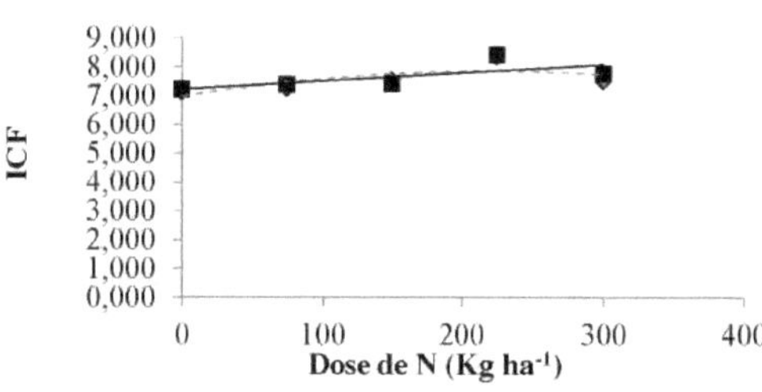
ICF clorofila b - 20 DAT
9,000
8,000
7,000
6,000
5,000
4,000
3,000
2,000
1,000
0,000
ICF
0
100
200
300
400
Dose de N (Kg ha-1)
Chl b (s/ sombra)
y = -2E-05x2 + 0,008x + 6,9713
R² = 0,5396
Chl b (c/ sombra)
y = 0,0028x + 7,2262
R² = 0,4848

ICF clorofila b - 40 DAT

ICF

12,000
10,000
8,000
6,000
4,000
2,000
0,000

0 100 200 300 400

Dose de N (Kg ha^{-1})

◆ Chl b (s/ sombra) $y = 6E\text{-}08x^2 + 0{,}0127x + 6{,}3147$
$R^2 = 0{,}9825$

■ Chl b (c/ sombra) $y = 0{,}0133x + 6{,}3614$
$R^2 = 0{,}9928$

ICF clorofila b - 60 DAT

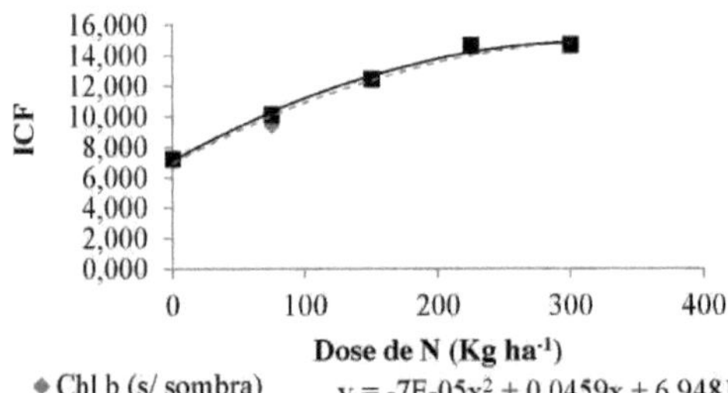

◆ Chl b (s/ sombra) $y = -7E\text{-}05x^2 + 0{,}0459x + 6{,}9481$
$R^2 = 0{,}981$

■ Chl b (c/ sombra) $y = -7E\text{-}05x^2 + 0{,}0482x + 7{,}1154$
$R^2 = 0{,}9916$

ICF clorofila b - 80 DAT

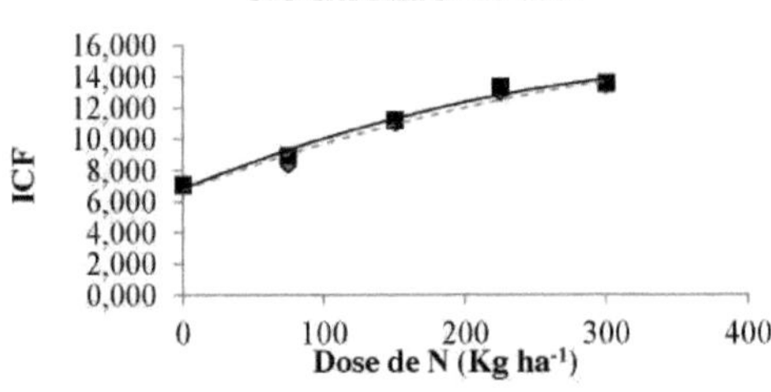

◆ Chl b (s/ sombra) $y = -3E\text{-}05x^2 + 0{,}0324x + 6{,}7509$
$R^2 = 0{,}9724$

■ Chl b (c/ sombra) $y = -4E\text{-}05x^2 + 0{,}0359x + 6{,}8397$
$R^2 = 0{,}9817$

ICF clorofila total - 20 DAT

ICF

40,000
35,000
30,000
25,000
20,000

0 100 200 300 400

Dose de N (Kg ha-1)

◆ Chl Total (s/ sombra) $y = -5E\text{-}05x^2 + 0,0274x + 30,285$ $R^2 = 0,9973$

■ Chl Total (c/ sombra) $y = 0,0149x + 31,103$ $R^2 = 0,7936$

ICF clorofila total - 40 DAT

ICF

40,000
35,000
30,000
25,000
20,000

0 100 200 300 400

Dose de N (Kg ha-1)

◆ Chl Total (s/ sombra) $y = -6E\text{-}05x^2 + 0,0545x + 25,761$ $R^2 = 0,9871$

■ Chl Total (c/ sombra) $y = 0,0372x + 27,04$ $R^2 = 0,986$

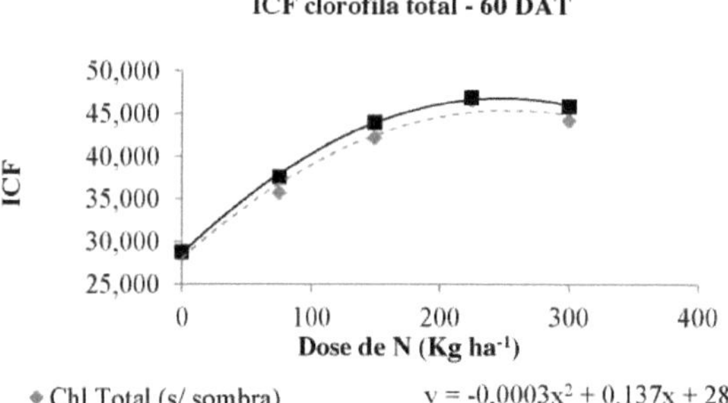

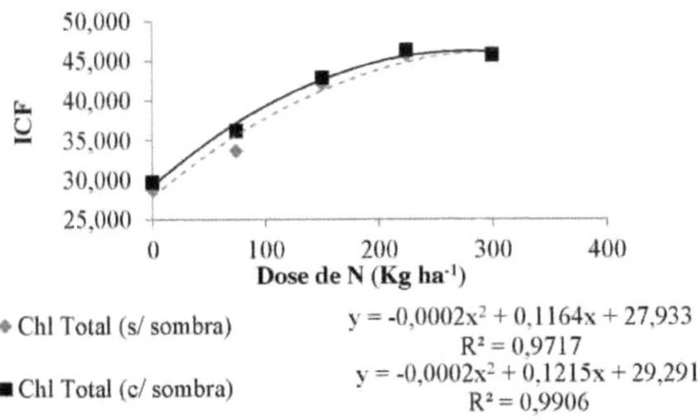

Figure 17 - ICF (Falker chlorophyll index) assessment of chlorophyll *a b* and total when subjected to increasing doses of nitrogen in two growing environments (with shading and without shading). Santa Maria, RS, 2017.

5.5 DENSITY AND STOMATAL DIAMETER

5.5.1 Stomatal density

According to the results of the analysis of variance, shown in Figure 18, there was a significant effect ($p \leq 0.05$) of the treatments (nitrogen doses and environment) on the variable stomata density (n°/mm^2) only in the shaded environment. The results obtained from the regression analysis (Figure 18) show that, in the shaded environment (significant effect), the stomata density evaluation fitted a quadratic function.

Tomato plants have amphistomatous leaves (with stomata on both sides). In the shaded environment, the number of stomata was higher in the treatment without nitrogen (138.0 and 42.86/mm^2 , on the abaxial and adaxial sides of the leaflets, respectively), with the values decreasing as the doses of the nutrient were applied and increased (Figure 18).

As a result, the number of stomata per mm2 decreased by an average of 30% in the leaflets of the plants grown with the highest dose of nitrogen (300 kg ha^{-1} of N), where this treatment had 97.138 and 18.32 stomata/mm^2 , on the abaxial and adaxial sides of the leaflets, respectively, when compared to the treatment without N. A plant's photosynthetic capacity is generally proportional to its nitrogen content (BOLTON & BROWN, 1980) and the lack of nitrogen associated with less light can lead to a high number of stomata (Figure 18).

Figure 18 shows that in the environment without shading, the number of stomata on both faces (adaxial and abaxial) was not significant, suggesting that the application of different doses of nitrogen does not influence the analyzed variable when the tomato plants are not shaded.

In relation to the environments, the number of stomata/mm2 on the leaflets of the tomato plants that were not shaded showed the highest values on average (58.62 and 127.0 on the adaxial and abaxial faces, respectively). These results corroborate those found by Santiago et al., 2001, where the researchers studying the leaf anatomy of long pepper under different light conditions concluded that, in the brighter environment, there was a higher stomatal index in the leaves of the plants. According to Larcher (2000), sunny leaves have higher stomatal densities than shady leaves. Increased stomatal indices are a common occurrence in sunny leaves compared to shady ones (NASCIMENTO et al., 2005; LIMA JUNIOR et al., 2006) and are generally associated with the smaller size of the stomata, which guarantees the supply of CO_2 needed for photosynthesis without excessive water loss through transpiration (MELO, 2007).

In both environments (with and without shading) and in all doses of nitrogen applied, the abaxial side of the leaflets had the highest number of stomata/mm^2 . A study carried out by Meyer et al. (1973), analyzing the distribution of the number of stomata in various plant species of agricultural interest, concluded that the tomato crop has more stomata in the lower epidermis (abaxial) than in the upper epidermis (adaxial), with averages of 13,000 and 1,200 stomata/cm^2 , respectively. Larcher (2000) states that the number of stomata is a specific characteristic of each species and can be altered as a result of adaptations to environmental conditions.

5.5.2 Stomatal diameter and functionality

The variables equatorial stomata diameter (ED), polar stomata diameter (PD) and stomatal functionality (FUNC) showed a significant effect ($p \leq 0.05$) in the environment where the plants were shaded, according to the results of the analysis of variance (Figure 18). In the environment without shading, only the equatorial diameter of the stomata (DE) and stomatal functionality (FUNC) variables showed a significant effect ($p \leq 0.05$). Through the results obtained by regression analysis (Figure 18), the variables that showed significant effects were fitted to quadratic functions.

The ratio between the polar diameter (PD) and equatorial diameter (ED) of the stomata is directly related to their functionality, and the higher the ratio, the more functional the stomata (CASTRO et al., 2009). Table 4 shows that the highest values of stomatal functionality (FUNC) in the two environments were related to the 150 kg ha^{-1} dose of N, with 1.757 in the shaded environment and 1.653 in the unshaded environment.

With regard to the size of the stomata, Nascimento et al. (2005); Lima Junior et al. (2006), state that the larger size of the stomata is a common characteristic of shade plants, due to the lower demand for water compared to plants that are more exposed to solar radiation, and therefore do not need to reduce the size of their stomata. According to Dickison (2000), the reduction in stomata size under high light conditions is considered a protective adaptation against dehydration. Figure 18 shows that the average polar diameters were 20.300 and 19.167 μm in the shaded and unshaded environments, respectively, and the average equatorial diameters were 12.833 and 14.800 μm. Castro et al. (2009) state that an increase in the polar diameter and a reduction in the equatorial diameter allow for a more elliptical shape of the stomata, leading to greater functionality. More elliptical stomata can be more functional, reducing plant transpiration (CASTRO et al., 2009).

With regard to stomatal functionality, the shaded environment showed the highest DP/DE ratio values, with an average of 1.589 compared to the unshaded environment (1.470). According to Kahn et al. (2003), changes in the shape of stomata affect their functionality, with the elliptical shape being characteristic of functional stomata, while the spherical shape is often associated with stomata with low functionality and both shapes are obtained according to the greater or lesser equatorial diameter and the DP/DE ratio. Variations in the size and frequency of stomata show the ability of plants to rearrange these epidermal structures in response to environmental changes, making stomata more active in gas exchange and adequate transpiration (ROSSATTO et al.,

2009).

The biometric and physiological results observed in this study corroborate the information provided by Taiz and Zeiger (2009), in which a reduction in stomatal functionality can cause damage to plants, such as a reduction in the availability of substrate (CO_2) for photosynthetic activity and consequently a reduction in their development, as observed for the treatments with shading and a low dose of N (Figure 18).

Table 4 - Evaluation of the equatorial (ED) and polar (PD) diameter of the stomata (ED) and stomatal functionality (FUNC) of tomato plants subjected to increasing doses of nitrogen in two growing environments (with shading and without shading). Santa Maria, RS, 2017.

Treatments '	With[1] DE	Without[2] DE	With DP	Without DP	With FUNC	No FUNC
0 kg ha^{-1} N	14,167	14,333	19,167	21,333	1,353	1,488
75 kg ha^{-1} N	13,000	13,333	20,167	17,000	1,551	1,275
150 kg ha^{-1} N	12,333	12,500	21,667	20,667	1,757	1,653
225 kg ha^{-1} N	12,667	11,833	20,667	17,333	1,632	1,465
300 kg ha^{-1} N	12,000	22,000	19,833	19,500	1,653	0,886

[1] Treatments applied to tomato plants in an unshaded environment.

[2] Treatments applied to tomato plants in a shaded environment.

*Means not followed by the same letters in the columns differ by Tukey's test at 5% probability of error.

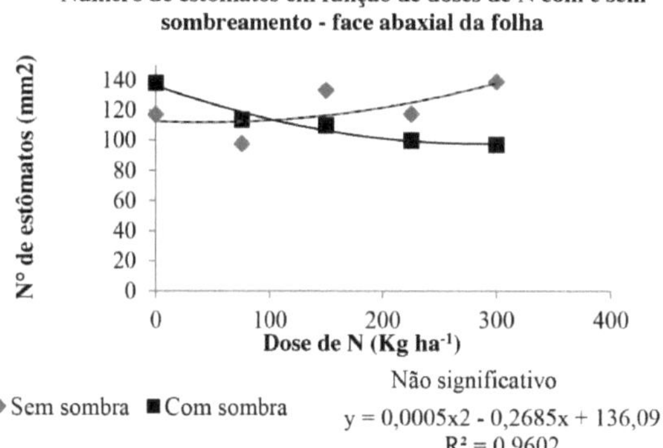

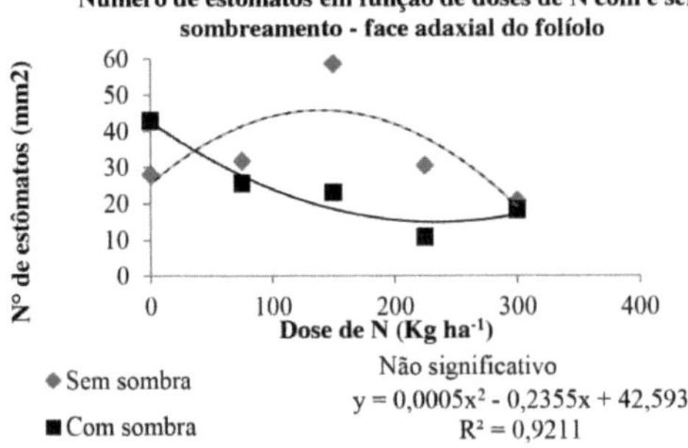
Número de estomatos em função de doses de N com e sem sombreamento - face adaxial do folíolo
N° de estômatos (mm2)
60
50
40
30
20
10
0
0
100
200
300
400
Dose de N (Kg ha-1)
◆ Sem sombra
■ Com sombra
Não significativo
y = 0,0005x² - 0,2355x + 42,593
R² = 0,9211

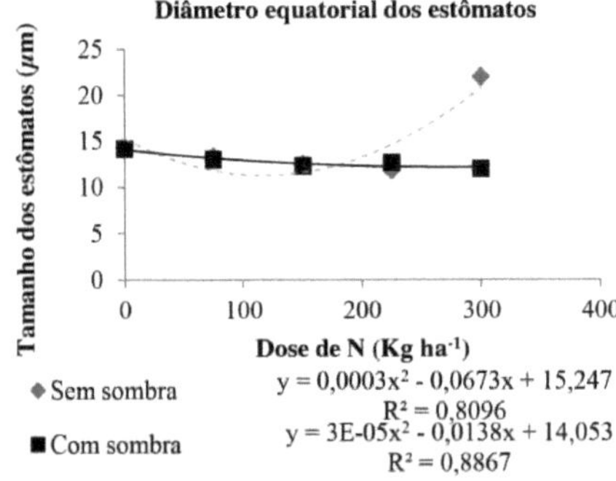
Diâmetro equatorial dos estômatos
Tamanho dos estômatos (μm)
25
20
15
10
5
0
0
100
200
300
400
Dose de N (Kg ha-1)
◆ Sem sombra
■ Com sombra
y = 0,0003x² - 0,0673x + 15,247
R² = 0,8096
y = 3E-05x² - 0,0138x + 14,053
R² = 0,8867

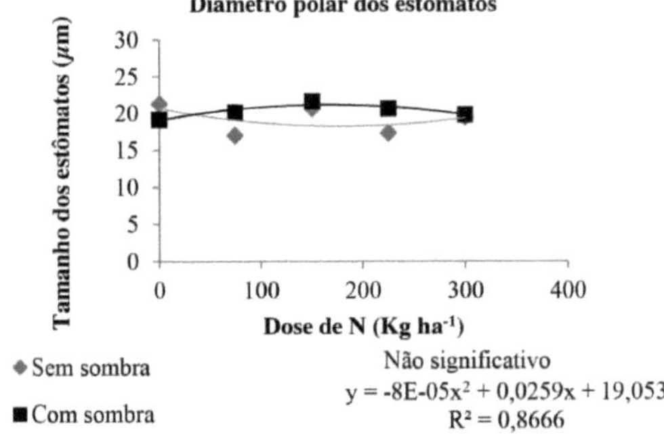
Diâmetro polar dos estômatos
Tamanho dos estômatos (μm)
30
25
20
15
10
5
0
0
100
200
300
400
Dose de N (Kg ha-1)
◆ Sem sombra
■ Com sombra
Não significativo
y = -8E-05x² + 0,0259x + 19,053
R² = 0,8666

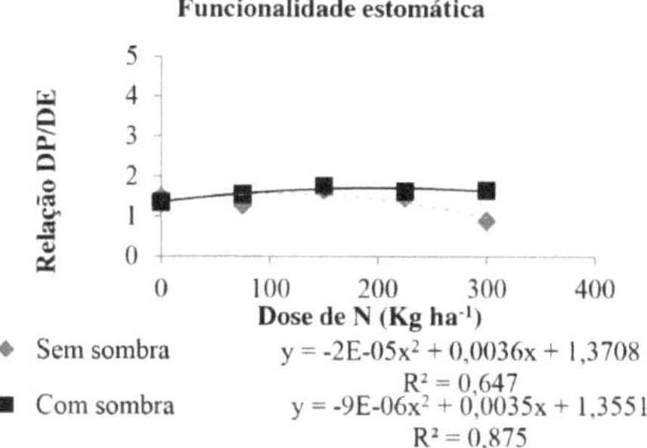

Figure 18 - Evaluation of the number/mm^2 and dimensions (μm) of the stomata of the leaflets of tomato plants subjected to increasing doses of nitrogen in two environments (with shading and without shading). Santa Maria, RS, 2017.

5.6 AERIAL PART DRY MASS

The dry mass of the aerial part of the plant is commonly used to assess the efficiency of nitrogen fertilization in crops, as it shows the accumulation of plant biomass due to the greater production of amino acids and carbon assimilates from photosynthesis, provided by the supply and absorption of nitrogen in the plant (PORTO, 2013). According to the results of the analysis of variance (Figure 21), there was a significant effect ($p \leq 0.05$) of the treatments (nitrogen doses and environment) on the variable dry mass of the aerial part of the tomato plants. The results obtained from the regression analysis (Figure 21) show that the variable fitted a quadratic function.

The dry matter of the aerial part remaining after the last fruit harvest increased according to the doses of nitrogen in the environments with and without shading, reaching maximum values of 26.204 and 28.585 g/plant, respectively. Ferreira et al., 2003, observed an increase in the dry matter of the aerial part of tomato plants as the doses of nitrogen applied increased. The treatments without N application resulted in the lowest aerial part dry matter values, 10.167 and 11.997 g, in the environments with and without shading, respectively.

The maximum values found in the evaluations carried out on the dry matter of the aerial part were observed in the treatment with 202.62 kg ha^1 of N in the unshaded environment and 209.62 kg ha^{-1} in the shaded environment.) After this treatment, PA dry matter values decreased as a result of the excessive doses of N applied. These results corroborate those found by Porto (2013), where the application of high doses of nitrogen can lead to toxic doses, reducing plant growth.

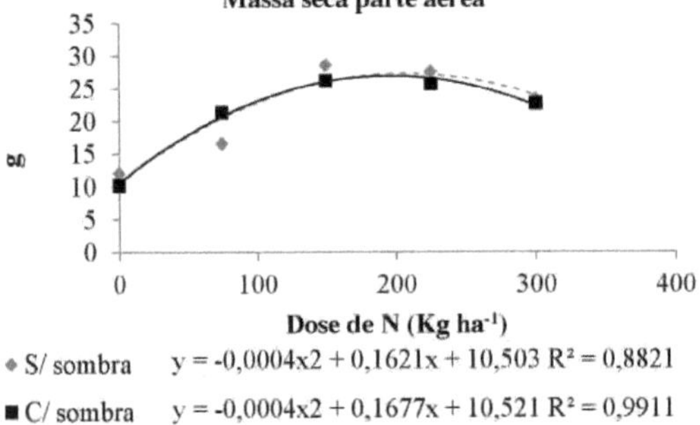

Figure 19 - Evaluation of the dry mass of the aerial part of tomato plants subjected to increasing doses of nitrogen in two environments (with shading and without shading). Santa Maria, RS, 2017.

5.7 TOTAL FRUIT PRODUCTION

The evaluations carried out on the variables number of fruits produced per treatment, average fruit weight (g) and fruit diameter and length (cm) were significantly affected ($p \leq 0.05$) by the increasing doses of nitrogen applied and by the environment (with and without shading) in which the tomato plants were located (Figure 20). In the evaluation of the number of fruits produced in the unshaded environment, the evaluations showed a quadratic behavior as a function of the increase in the dose of N and in relation to the environment.

In both environments (with and without shading), the treatment without nitrogen application had the lowest values for the number of tomato fruits, with 19 and 17 fruits, respectively. This result is in line with Mehmood et al (2012) who, when evaluating the response of tomato cultivars by varying the level of N, observed that the lowest number of fruits per plant (21.82) was produced by the treatment without N application. The other variables analyzed (fruit weight, diameter and length) also showed the lowest values found in the treatment without nitrogen application. These characteristics are extremely influenced by nitrogen levels, where a deficiency of this nutrient in the tomato crop leads to lower growth, number and size of fruit (SAINJU et al., 2003).

With regard to the shaded environment, the variables number of fruits, average fruit weight (g), fruit length and diameter (cm) showed a quadratic function. It was observed that the treatment with the dose of 168.8 kg ha^{-1} of N applied had the highest value of total fruit, with 54 tomato fruit; the dose of 166.8 kg hà$^{-1}$ of N had an average of 102.82 g fruit^{1} ; the dose of 150 kg hà$^{-1}$ of N, 5.63 cm fruit length^{-1} and 4.57 cm fruit diameter^{-1} , at the dose of 150 kg hà$^{-1}$ of N (Figure 20).

The reduction in values found with doses of nitrogen higher than 150 kg ha^{-1} is attributable to excess N in the plant, causing toxicity in the plant which results in a reduction in yield (Figure 20). Despite the reduction found, it is well known that tomato yields are influenced by N, among other benefits, because it increases the absorption of other nutrients (AMAN & RAB, 2013).

In the unshaded environment, the number of tomato fruits fitted a linear equation as the dose of nitrogen applied increased, with the highest dose (300 kg ha^{-1} of N) producing a total of 78 fruits. Ferreira et al (2000) and Silva et al (2003) also observed that applying nitrogen increased tomato fruit production. With regard to the other variables analyzed, average fruit weight (g), fruit length and diameter (cm) showed a reduction in their values as the doses of N applied increased. According to the results shown in Figure 20, the highest values were found with an application of 144 kg ha^{-1} of N, obtaining an average of 130.93 g fruit^{-1} 6.24 cm fruit length^{-1} and 5.32 cm fruit diameter^{-1} , with an application of 150 kg hà-1 of N. The results for average fruit weight are higher than those found by Silva et al (2003) who observed an average of 102.5 g per table tomato fruit.

As can be seen in Figure 20, the environment without shading showed the highest values for the variables number of fruits produced per treatment, average fruit weight (g) and fruit diameter and length (cm) when compared to the environment with shading. These results corroborate those found by Santi (2014), where the average weight of the fruit was significantly higher in the environment with only polyethylene, showing the greater efficiency of the covering material in generating more suitable microclimatic conditions, in terms of the quality of the available radiant energy.

The lower number of fruits per treatment in the shaded environment can be attributed to the fact that some of the plants in this environment were not in the flowering stage, resulting in lower fruit production in this environment. Although the tomato crop is considered to be insensitive to photoperiod, it does have light requirements. According to Reis et al. (2013), increased radiation can increase the production of photoassimilates and their availability for plant growth and fruit production. Therefore, solar radiation influences plant development, from the time of flowering with effects on inflorescence initiation and the rate of flower development to the growth of the stem system in summer (AGEITEC, 2014).

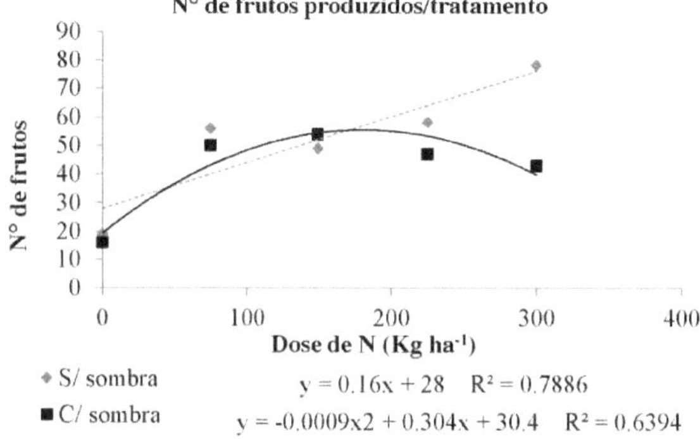
N° de frutos produzidos/tratamento
90
80
70
60
50
40
30
20
10
0
N° de frutos
0
100
200
300
400
Dose de N (Kg ha-1)
S/ sombra
C/ sombra
y = 0.16x + 28 R² = 0.7886
y = -0.0009x2 + 0.304x + 30.4 R² = 0.6394

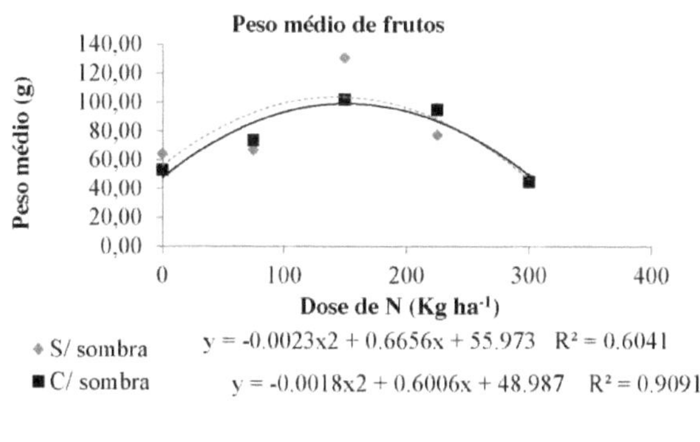
Peso médio de frutos
140,00
120,00
100,00
80,00
60,00
40,00
20,00
0,00
Peso médio (g)
0
100
200
300
400
Dose de N (Kg ha-1)
S/ sombra
C/ sombra
y = -0.0023x2 + 0.6656x + 55.973 R² = 0.6041
y = -0.0018x2 + 0.6006x + 48.987 R² = 0.9091

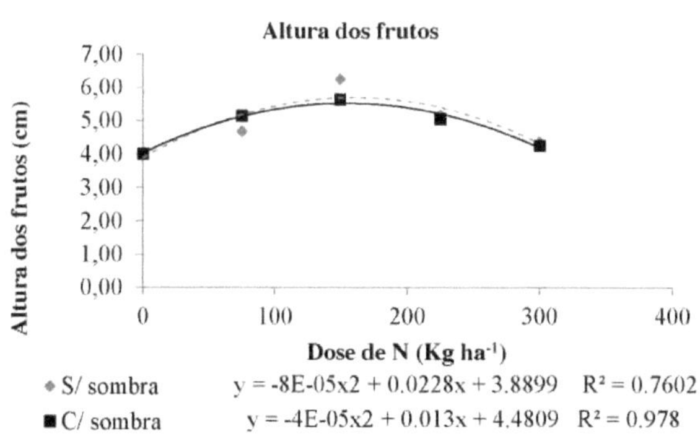
Altura dos frutos
7,00
6,00
5,00
4,00
3,00
2,00
1,00
0,00
Altura dos frutos (cm)
0
100
200
300
400
Dose de N (Kg ha-1)
S/ sombra
C/ sombra
y = -8E-05x2 + 0.0228x + 3.8899 R² = 0.7602
y = -4E-05x2 + 0.013x + 4.4809 R² = 0.978

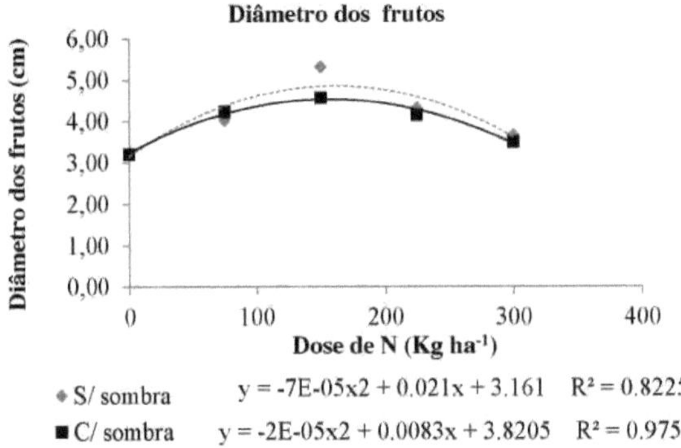

Figure 20 - Evaluation of the number of fruits, weight (g), average diameter and length (cm) of the fruits when subjected to increasing doses of nitrogen in two environments (with shading and without shading). Santa Maria, RS, 2017.

5.8 FRUIT QUALITY

According to the results of the analysis of variance, shown in Figure 21, there was a significant effect ($p \leq 0.05$) of the treatments (nitrogen doses and environment) on the variables fruit pH, titratable acidity (g citric acid/100 ml), soluble solids (°Brix) and soluble solids/titratable acidity ratio (SS/AT). The results of the regression analysis (Figure 21) show that all the variables analyzed fitted a quadratic function.

5.8.1 pH

The pH of the tomato fruit was altered by the doses of nitrogen and the environment the plants were in (with and without shading). In the shaded environment, the fruit reached average pH values of 5.17 to 5.33, in the treatments corresponding to doses of 0 and 225 kg ha^{-1} of N, respectively (Figure 21 A).

In the unshaded environment, the average pH values ranged from 4.27 to 4.80 in the 0 and 225 kg ha^{-1} N treatments, respectively (Figure 20 A). As a result, as the dose of N increased, the pH values rose regardless of the environment to which the plants were subjected (with or without shading). These results corroborate those found by Assunçao (2016), who found that increasing the dose of nitrogen led to a decrease in the average pH value found for the flesh of tomato fruit.

The recommended nitrogen dose for tomato growing based on soil analysis and tomato needs (150 kg ha^{-1} of N) showed average values of 5.27 and 4.53 in the shaded and unshaded environments, respectively (Figure 21 A). The pH of tomato fruit destined for industrial processing must vary between 4.0 and 4.5 to inhibit the growth of bacteria (JONES J., 1999). Thus, the flesh of the tomato fruit in the unshaded environment had pH values closest to the ideal values, reducing the risk of contamination by microorganisms.

5.8.2 Titratable acidity

Evaluating the titratable acidity content in the pulp of tomato fruits under different nitrogen doses and different environments showed changes in the values for these two variables. In the shaded environment, the average values were 0.2192% citric acid and in the unshaded environment, 0.2274% (Figure 21 B). This differs from the results observed by Sjostrom & Rosa (1978) and Ritzinger et al. (1989), who reported an increase in acidity in fruit harvested under lower temperatures and lower radiation. High levels of acid in the pulp are an important characteristic when it comes to processing, as it is interesting for the fruit to have high acidity, as this would reduce the need to add acidifiers to the juice (NASCIMENTO, 1996).

Citric acid is considered to be the most present acid in the fruit and begins to accumulate soon after the fruit is formed, quickly reaching its maximum value. Nutritional conditions and temperature are the factors that most influence the accumulation of this acid. After reaching its maximum value, the concentration decreases due to the absorption of water by the fruit, thus diluting the acid present (CAVICHIOLI et al., 2008). According to Rasmussen et al. (1996) and Albrigo (1992), the higher the temperature during fruit ripening, the greater the decrease in acid concentration. This was not observed in this experiment due to the difference in temperature values found in the two environments being only 1.91% and considered low for this factor to occur (20.5 and 20.9°C, respectively in the environments with and without shading).

The different doses of nitrogen applied altered the titratable acidity of the fruit pulp. In the shaded environment, the maximum value found was 0.2323% citric acid, at a dose of 214 kg ha^{-1} of N. In the unshaded environment, the treatment with 150 kg ha^{-1} of N showed the highest value, 0.2423%. Nitrogen doses higher than 214 kg ha^{-1} and 150 kg ha^{-1} (with and without shading, respectively) negatively affected the quality of the fruit in terms of acidity, since the results for % citric acid were lower (Figure 21 B). According to Jones Junior (1999), excessive doses of nitrogen negatively affect the quality of tomato fruit in terms of acidity, as the higher the acidity, the better the taste of the tomato.

Oberly et al. (2002), when evaluating the effect of nitrogen fertilization on the quality of four tomato cultivars, observed a difference between the control and the nitrogen doses, where the titratable acidity values of the fruit were higher when the soil received higher doses of nitrogen, confirming the results found in this study. Evaluating the impact on the production and quality of tomato fruit, Bérnard et al. (2009) found a 10% reduction in titratable acidity in response to a reduction in the nitrogen supply. In this way, the doses of 214 and 150 kg ha^{-1} in the environments with and without shading, respectively, showed the highest values of citric acid percentage (0.2323 and 0.2423, with and without shading, respectively) and consequently, better quality in terms of titratable acidity.

5.8.3 Soluble solids

The percentage of soluble solids present in fruit pulp is related to taste and is represented by °Brix. Most tomato cultivars produce fruit with °Brix ranging from 5.0 to 7.0 (FERREIRA et al., 2006). In this study, the average values were 4.89 and 4.93 in the shaded and unshaded environments, respectively, indicating that the environmental conditions given to the tomato plants led to a difference in the soluble solids content (Figure 21 C). Dhillon et al. (1990) state that high average temperatures and high luminosity increase the soluble solids content, due to the greater photosynthetic activity and the greater accumulation of carbohydrates in the fruit.

The different doses of nitrogen applied altered the °Brix levels found in both environments. The highest °Brix values were 5.93 (268 kg ha^{-1} of N) and 6.00 (236 kg ha^{-1} of N) in the environments with and without shading, respectively. The doses of nitrogen applied thereafter reduced the °Brix content of the tomato fruit pulp. In the work carried out by Anaç et al. (1994), when nitrogen was applied, the soluble solids values were constant up to a rate of 240 kg ha^{-1} , after which there was a reduction in °Brix. In this same study, the use of nitrogen resulted in a lower value for this variable (5.84) when compared to the present study (Figure 21 C).

Nitrogen plays an important role in the biosynthesis of sugars in the leaves, which are translocated to the fruit and can increase the concentration of soluble solids in the fruit (FERREIRA et al., 2006). In addition to genetic effects, other factors such as temperature, water, fertilizer and light determine the plant's level of photosynthesis and, consequently, the amount of sugars and dry matter in the fruit (PIERRO, 2002). For tomato fruits that are used in processing, an increase in the soluble solids content has a major influence on industrial yield, since the higher the °Brix, the higher the yield and the less energy is needed to concentrate the pulp (SILVA and GIORDANO, 2000).

5.7.4 Soluble solids/titratable acidity (SS/AT)

Based on the results found for the soluble solids content and titratable acidity variables, the soluble solids/titratable acidity (SS/AT) ratio is calculated (Figure 21 D). According to Jones Júnior (1999), the higher the acidity and sugar content, the better the taste of the tomato. According to Kader et al. (1978), high quality fruit should have a SS/AT ratio greater than 10. A high SS/AT ratio indicates an excellent combination of sugar and acid, which are related to a mild flavor, while low values are related to acid and a worse fruit flavor (PACHECO et al., 1997).

Figure 21 D, which shows the SS/AT ratio, shows that the dose of 245 kg ha^{-1} of N had a ratio value of 25.5882 in the shaded environment and in the unshaded environment the dose of 194 kg $hà^{-1}$ of N had a value of 24.7661. The lowest SS/AT ratios were 15.5743 and 16.4312 in the shaded and unshaded environments, respectively, and in the treatments without nitrogen. This shows that, as the dose of nitrogen applied increases, the SS/AT ratio grows. These results corroborate those found by Kobryn & Hallmann (2005), where there was an increase in this ratio when nitrogen doses were increased.

The results show that the doses of nitrogen and the two environments produced fruit of adequate quality, as their values were above 10. As with acidity and soluble solids content, the difference in the SS/AT ratio of fruit can be influenced by cultivars, stages of ripeness, management, fertilization, irrigation and soil composition (FELTRIN et al., 2002).

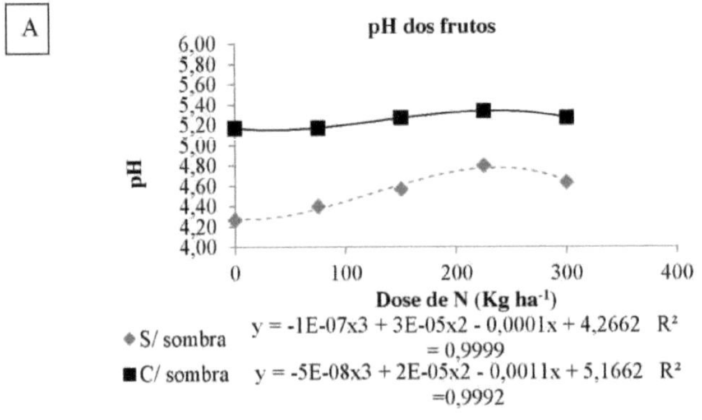

B

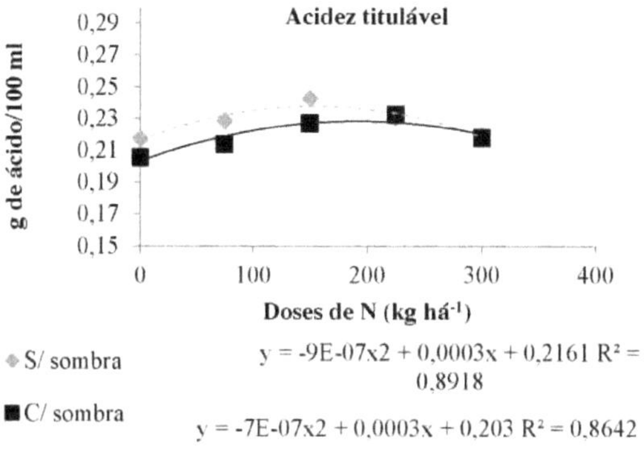

C

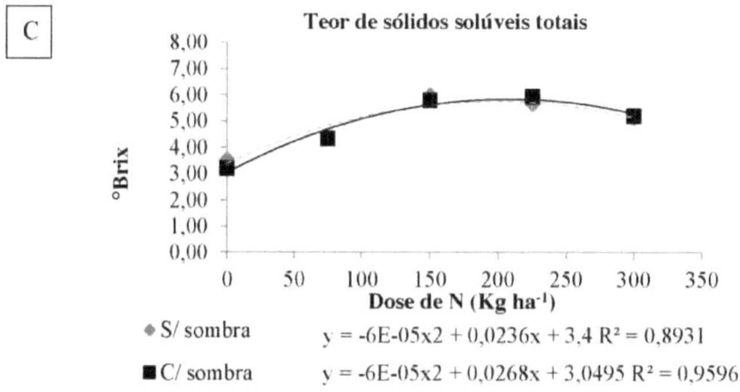

D

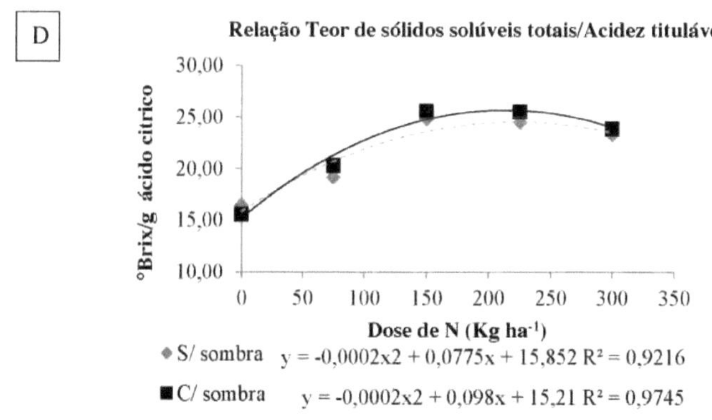

Figure 21 - Evaluation of pH, titratable acidity (TA), soluble solids (SS) and SS/TA ratio of fruit when subjected to increasing doses of nitrogen in two environments (with shading and without shading). Santa Maria, RS, 2017.

6 CONCLUSION

Of the parameters evaluated in the tomato crop, it was found that productivity and fruit quality responded satisfactorily up to the dose of 225 kg ha-1 of N in both environments (with and without shading), 50% more than the recommended dose according to the soil analysis and recommendation for the crop. The doses prior to this (0, 75 and 150 kg ha^{-1} of N) and after this (300 kg ha^{-1} of N) showed lower growth, development, productivity and fruit quality.

The environment without a 50% black shading screen provided better agronomic performance for tomato plants grown in a protected environment, giving higher yields and tomato fruit quality compared to the environment without a 50% black screen.

REFERENCES

ADAMS, P. Mineral nutrition. In: Atherton, J. G.; Rudich, J. (ed.). The tomato crop: A scientific basis for improvement. London/ New York: **Chapman and Hall**, 1986. ch.7, p.281-334.

AGEITEC. Embrapa Agency for Technological Information. Available at: http://www.agencia.cnptia.embrapa.br. Accessed on: 22/01/2017.

ALBRIGO, G. Environmental influences on citrus fruit development. In: INTERNATIONAL CITRUS SEMINAR - PHYSIOLOGY, 2., 1992, Bebedouro-SP. **Proceedings...** Campinas: Cargill Foundation, 1992. p.100-105.

ALVARENGA, M. A. R. Tomato, field, greenhouse and hydroponic production: **Botanical origin and plant description**. Lavras-MG, 2013, 455p.

AMAN, S.; RAB, A. Response of tomato to nitrogen levels with or without humic acid. **Sarhad Journal of Agriculture**, v. 29, p. 181-186, 2013.

ANAÇ, D. ERIUCE, N.; KILINÇ, R. Effect of N, P, K fertilizer levels on yield and quality properties of processing tomatoes in Turkey. **Acta Horticulturae,** n. 376, p. 243-250. 1994.

ANDRIOLO, J. L. Physiology of vegetable production in a protected environment. **Horticultura Brasileira**, Brasilia, v. 18, p. 26-33, 2000. Supplement.

ARAUJO, C.; FONTES, P.C.R.; SEDIYAMA, C.S.; COELHO, M.B. Criteria for determining the dose of nitrogen to be applied to tomato plants in a protected environment. **Horticultura Brasileira**, v. 25, p. 327-332, 2007.

ARGERICH, C.; TROILO, L. **Manual de buenas pràcticas Agricolas en la cadena de tomate:** Aspectos fisiologicos del cultivo de tomate, Buenos Aires - ARG, 2011, 262p.

ASHKEY, D.A.; DOS, B.D.; VENNETT, O.L. A method of determining leaf area in cotton. **Agronomy Journal,** Madison, v.25, p.484-585, 1963.

ATARASSI, R. T. Modeling the energy balance of the pepper crop canopy in a greenhouse. ESALQ/USP: Piracicaba, 2004. 70p. PhD **thesis**.

BECKMANN, M.Z.; DUARTE, G.R.B.; PAULA, V.A.; MENDEZ, M.E.G.; PEIL, R.M.N. Solar radiation in a protected environment cultivated with tomato in the summer-autumn seasons of Rio Grande do Sul. **Ciência Rural**, Santa Maria, v.36, n.1, p.86-92, jan-feb, 2006.

BÉNARD, C.; GAUTIER, H.; BOURGAUD, F.; GRASSELLY, D.; NAVEZ, B.; CARISVAYRAT, C.; WEISS, M.; GENARD, M. Effect of low nitrogen supply on tomato (*Solanum lycopersicum*) fruit yield and quality with special emphasis on sugar, acids ascorbate, carotenoids and phenolic compounds.

Journal of Agricultural and Food Chemistry, v. 57, p. 4112-4123, 2009.

BOLTON, J.K.; BROWN, R.H. Photosynthesis of grass species differing in carbon dioxide fication pathways V. Response of panicum maximum, panicum milioides, and tal-fescue (Festuca arundinacea) to nitrogen nutrition. **Plant Physiology**, v.66, p. 97-100, 1980.

BORGOGNONE, D.; COLLA, G.; ROUPHAEL, Y.; CARDARELLI, M.; REA, E.; SCHWARZ, D. Effect of nitrogen form and nutrient solution pH on growth and mineral composition of self-grafted and grafted tomatoes. **Scientia Horticulturae**, v. 149, p. 61-69, 2013.

BRICKELL, C.D.; BAUM, B.R.; HETTERSCHEID, W.L.A.; LESLIE, A.C., MCNEILL, J.; TREHANE, P.; VRUGTMAN, F.; WIERSEMA, J.H.

International code of nomenclature of cultivated plants. **Acta Horticulturae**, v.647, p.1-123, 2004. http://www.bgbm.fu-berlin.de/iapt/nomenclature/code/SaintLouis/0001ICSLContents.htm. (October 22, 2016).

BRITTO, D.T.; GLASS, A.D.M.; KRONZUCKER, H.J.; SIDDIQI, M.Y. Cytosolic concentrations and transmembrane fluxes of NH4+/NH3. An evaluation of recent proposals. **Plant Physiology**, v. 125, p. 523-526, 2001.

CAMARGO, F. P.; FILHO, W. P. C. (2008). Table tomato production in Brazil, 1990-

2006: Contribution of area and productivity. **Horticultura Brasileira**, v.26, n.2, p.S1018-S1021.

CAMPOS, A.R.F. Organic and mineral fertilization on the production characteristics of the Santa Cruz tomato cultivar in a protected environment. Agronomy **Course Conclusion Paper**. Federal University of Paraiba, Areia- PB, 2013.

CAMPOS, M.A.A.; UCHIDA, T. Influence of shading on the growth of seedlings of three Amazonian species**. Pesq. Agropecu. Bras**., Brasilia, v. 37, n. 3, p. 281-288, mar. 2002.

CAMPOS, M.A.A.; UCHIDA, T. Influence of shading on the growth of seedlings of three Amazonian species**. Pesq. Agropecu. Bras**., Brasilia, v. 37, n. 3, p. 281-288, mar. 2002.

CARRIJO, O. A.; VIDAL, M. C.; REIS, N. V. B. DOS; SOUZA, R. B. DE; MAKISHIMA, N. Tomato productivity in different substrates and greenhouse models**.** **Horticultura Brasileira** , Jan./Mar. 2004. v. 22, n. 1.

CAVICHIOLI, J.C.; RUGGIERO, C.; VOLPE, C.A.; PAULO, E.M.;

FAGUNDES, J.L.; KASAI, F.S. Flowering and fruiting of yellow passion fruit submitted to artificial lighting, irrigation and shading. **Revista Brasileira de Fruticultura**, Jaboticabal, v.28, n.1, p.92-96, 2006.

CERMENO, Z.S. **Greenhouses: Installation and management**. Lisbon: Litexa, 1990. 355p.

CRAWFORD, N. M., and FORBE, B. J. 2002. Molecular and developmental biology of inorganic nitrogen nutrition. In: **The Arabidopsis Book**, Somerville, C. and Meyerowitz, E. eds. American Society of Plant Physiologists, Rockville, MD, pp. Doi/10.1199/tab.0011, http://www.aspb.org/publications/arabidopsis/.

CRAWFORD, N.M. Nitrate: nutrient and signal for plant growth. **The Plant Cell**, Rockville, v.7, p.859-868, 1995.

CREGAN, P.B., BERKUM, P. Genetics of nitrogen metabolism and physiological/biochemical selection for increased grain crop productivity. **Theoretical and Applied Genetics**, Heidelberg, v.67, p.97-111, 1984.

DICKISON, W.C. Integrative Pant Anatomy. USA, **Academic Press**. 533p. 2000.

DHILLON, B.S.; SINGH, S.N.; KUNDAL, G.S. Studies on the developmental physiology of guava fruit (Psodium guajava L.) II. Biochemical characters. **Horticultural Journal**, v.27, n.3-4, p.212-221, 1990.

DING, X.; GUO, Y.; NI, T.; KOKOT, S. A novel NIR spectroscopic method for rapid analyses of lycopene, total acid, sugar, phenols and antioxidant activity in dehydrated tomato samples. **Vibrational Spectroscopy**, v. 82, p. 1-9, 2016.

ELIA, A.; CONVERSA, G. Agronomic and physioological responses of tomato crop to nitrogen input. **European Journal of Agronimy**, v. 40, p. 64-74, 2012.

EPSTEIN E., and BLOOM, A.J. (2005). **Mineral Nutrition of Plant: Principles and Perspectives**, 2nd ed. Sinauer Associates, Sunderland, MA.

FAOSTAT (a) - FOOD AND AGRICULTURE ORGANIZATION OF THE UNITED NATIONS FAOSTAT (2013) - World Productivity. Available at: http : //www.fao. org/faostat/es/#data/QC Accessed on. 20/10/2016.

FAOSTAT (b) - FOOD AND AGRICULTURE ORGANIZATION OF THE UNITED NATIONS FAOSTAT (2013) - Food Supply. Available at: < http://faostat.fao.org/site/609/DesktopDefault.aspx?PageID=609#ancor>.

Accessed on. 20/10/2016.

FELTRIN, D. M.; LOURENÇÂO, A. L.; FURLANI, P. R.; CARVALHO, C. R. L. Effects of potassium sources on *Bemisia Tabaci* biotype B infestation and tomato fruit characteristics under protected environment. **Bragantia**, Campinas, v. 61, n. 1, p. 49-57, 2002.

FERREIRA, D. F. **Sisvar** - variance analysis system for balanced data. Lavras: UFLA, 1998. 19 p.

FERREIRA, M.M.M.; FERREIRA, G.B.; FONTES, P.C.R. Efficiency of tomato nitrogen fertilization in two growing seasons. **Revista Ceres**, v. 57, p. 263-273, 2010.

FERREIRA, M.M.M; FONTES, P.C.R. nitrogen indices in tomato leaves as a function of nitrogen and organic fertilization. **Revista Agro@mbiente**, v. 5, n. 2, p.106- 112, 2011.

FIGUEIREDO, G. Overview of production in protected environments. Casa da agricultura, protected environment production. 2011. Available at: http://www.asbraer.org.br/arquivos/bibl/56-ca-producao.pdf. Accessed on: 10 Dec 2016.

FILGUEIRA, F. A. R. **Manual de Olericultura:** Solanaceas II, Tomate: A mais universal das hortaliça, Sao Paulo-SP, Campus, 1982.

FILGUEIRA, F.A.R. **Novo Manual de Olericultura: agrotecnologia moderna na produçâo e comercializâo de hortaliças** - Viçosa, UFV, 2000.

FILGUEIRA, F. A. R. **Novo manual de Olericultura:** Solanaceas II, Tomate: A hortaliça cosmopolita, 3rd Edition, Viçosa-MG, Campus, 2008.

FLOSS, Elmar L. **Physiology of cultivated plants**: the study behind what you see. 5[a] edition, Passo Fundo, 2011, 734p.

FONTES, P. C.R.; PEREIRA, P. R. G. Nutriçao mineral do tomate para mesa In: **Informe Agropecuàrio**, Belo Horizonte, v. 24, n. 219, p. 27-34, 2003.

FONTES, P.C.R. Vegetable Production in a Protected Environment. **Informe Agropecuàrio**, Belo Horizonte, Sep/Dec. 1999. v.20, n.200/201, p.1-2.

FURLANI, A. M. C. Mineral Nutrition. In: KERBAUY G. B. (Org.). **Fisiologia Vegetal**. 1st ed. Rio de Janeiro: Guanabara Koogan, p. 40-75, 2004.

GARGANTINI, H.; BLANCO, H. G. Nutrient uptake by tomato plants. **Bragantia**, v. 56, n. 22, p. 693-714, 1963.

GHANEM, M.E.; MÂRTINEZ-ANDÛJAR, C.; ALBACETE, A.;

POSPISILOVÂ, H.; DODD, I.C.; PÉREZ-ALFOCEA, F.; LUTTS, S. Nitrogen forms alters hormonal balance in salt-treated tomato (*Solanum lycopersicum* L.). **Journal of Plant Growth Regulation**, v. 30, p. 144-157, 2011.

GIL PT; FONTES PCR; CECON PR; FERREIRA FA. 2002. SPAD index for the diagnosis of nitrogen status and for the prognosis of potato productivity. **Horticultura Brasileira** 20: 611-615.

GIORDANO, L. B.; RIBEIRO, C. S. da Botanical origin and chemical composition of the fruit. In: SILVA, Joao B. C. da; GIORDANO, L de B. (Org.) Tomate para o processamento industrial. Brasilia-DF. Embrapa Hortaliças. **Embrapa Communication for Technology Transfer.** p.36-59, 2000.

GOTO, R. Plasticulture in the tropics: a technical and economic evaluation.

Horticultura brasileira, Brasilia, v.15, p.163-165, 1997. Lecture. Supplement.

GUIL-GUERRERO, J.L.; REBOLLOSO-FUENTES, M.M. Nutrient composition and antioxidant activity of eight tomato (*Lycopersicum esculentum*) varieties. **Journal of Food Composition and Analysis**, v. 22, p. 123-129, 2009.

HAQUE, M.M.; HAMID, A.; BHUIYAN, N.I. Nutrient uptake and productivity as affected nitrogen and potassium application levels in maize/sweet potato intercropping system. Korean **Journal of Crop Science**, v. 46, p. 1-5, 2001.

HARVEY, M.; QUILLEY, S.; BEYNON, H. Exploring the tomato: transformations of nature, society and economy. Cheltenham, UK: Edward Elgar, 2002. 324p.

HERNÂNDEZ SUÂREZ, M.H.; RODRIGUEZ, E.M.R.; ROMERO, C.D. Mineral and trace element concentrations in cultivars of tomato. **Food Chemistry**, v. 104, p. 489-499, 2007.

HUETT, D.O.; DETTMANN, E.B. Effect of nitrogen on growth, fruit quality and nutrient uptake of tomatoes grown in sand culture. **Australian Journal of Experimental Agriculture**, v. 28, p. 391-399, 1988.

HUPPE, H.C., TURPIN, D.H. Integration of carbon and nitrogen metabolism in plant and algal cells. **Annual Review of Plant Physiology and Plant Molecular Biology**, Palo Alto, v.45, p.577-607, 1994.

IBGE (b) - BRAZILIAN INSTITUTE OF GEOGRAPHY AND STATISTICS. Report on the production of temporary crops from 2008 to 2010.

Available at: < http://seriesestatisticas.ibge.gov.br/lista_tema.aspx?op=0&no=1> Accessed on: 10/10/2016.

IBGE - Aggregate Data Bank. Available at: http://www.sidra.ibge.gov.br/bda/tabela/protabl.asp?c=1618&z=t&o=26&i=P. Accessed on: December 3, 2016.

IBGE - Systematic Survey of Agricultural Production. Available at: ftp://ftp.ibge.gov.br/Producao _Agricola/Levantamento_Sistematico_da_Produca o Agricola [mensal]/Comentarios/lspa 201501 comentarios.pdf. Accessed on: 02 Nov. 2016.

IGLESIAS, M.J.; GARCÌA-LÓPEZ, J.; COLLADOS-LUJAN, J.F.; LÓPEZ-

ORTIZ, F.; DiAZ, M.; TORESANO, F.; CAMACHO, F. Differential response to environmental and nutritional factors of high-quality tomato varieties. **Food Chemistry,** v. 176, p. 278-287, 2015.

ADOLFO LUTZ INSTITUTE. Analytical Standards of the Adolfo Lutz Institute.

Chemical and physical methods for food **analysis**. Volume 1. 3. ed., Sao Paulo, 2008. 533p.

JONES JÙNIOR JB. 1999. *Tomato plant culture: in the field, greenhouse and home garden.* Florida: CRC Press. 199p.

KADER, A. A. Postharvest technology of horticultural crops. Davis: **University of California**. p.535, 2002.

KANIsZEWsKI, s., ELKNER, K., RuMPEL, J. Effect of nitrogen fertilization and irrigation on yield nitrogen status in plants ans quality of fruits od firect seeded tomatoes. **Acta Horticulturae,** n. 200, p. 195-202, 1987.

KANIsZEWsKI, s., RuMPEL, J. The effect of nitrogen fertilization on the yield, nutrient status and quality of tomatoes under single and multiple harvest. **Biul Warzyw**. supplement, p. 19-29. 1983.

KOBRYN, J.; HALLMANN, E. The Effect of Nitrogen Fertilization on the Quality of Three Tomato Types Cultivated on Rockwool. **Acta Horticulturae**. v.691, p.56-58, 2005.

KROSS, R.K.; CAVALCANTI MARTA, M.E.R.M.; BRAGA, E.M. Influence of tomato (*Lycopersicon esculentum* L.) epidermis on mass transfer during osmotic treatment. Latin American Symposium on Food Science (SLACA), 4. Campinas, **Anais...** Campinas: UNICAMP, 2001.

KUMAR, M.; MEENA, M.L.; KUMAR, S.; MAJI, S.; KUMAR, D. Effect of nitogen, phosphorus and potassium fertilizers on the growth, yield and quality of tomato var. Azad t-6. **The Asian Journal of Horticulture**, v. 8, p. 616-619, 2013.

LARCHER, W. **Plant Ecophysiology**. Sao Carlos: RIMA Publishing House, 2000. 531 p.

LARSSON, C.M., INGEMARSSON, B. Molecular aspects of nitrate uptake in higher plants. In: WRAY, J.L., KINGHORN, J.R. Molecular and genetics aspects of nitrate assimilation. **Oxford: Oxford Science**, 1989. Chapt.1. p.3-14.

LIMA JUNIOR, E.C.; ALVARENGA, A.A.; CASTRO, E.M.; VIEIRA, C.V.;

BRASBOSA, J.P.R.A.D. Physioanatomical aspects of young plants of *Cupania vernalis* Camb. subjected to different levels of shading.

Revista Arvore, Viçosa, v.30, n.1, p.33-41, 2006.

LOOMIS, R. S., and CONNOR, D. J. (1992). **Crop Ecology: Productivity and Management in Agricultural Systems**. Cambridge University Press, Cambridge.

LOPES MC; STRIPARI PC. 1998. The tomato crop. In: GOTO R;

TIVELLI SW (ed). **Vegetable production in a protected environment**: subtropical conditions. Sao Paulo: UNESP. p. 257-319.

LOURES, J.L.; FONTES, P.C.R.; SEDIYAMA, M.A.N.; CASALI, V.W.D.; CARDOSO, A.A. Production and nutrient content of tomato plants grown in substrate containing pig manure. **Horticultura Brasileira**, Brasilia, v.16, n.1, p.50- 55, 1998.

MAIA, S. C. M. Use of a portable chlorophyll meter to determine nitrogen fertilization in bean cultivars. 2011. 86 f. **Dissertation** (Master's Degree in Agronomy)-Faculty of Agronomic Sciences, Universidade Estadual Paulista, Botucatu, 2011.

MALAVOLTA, E.; VITTI, G. C.; OLIVEIRA, S.A. de. Assessing the nutritional status of forage plants. In: Liming and fertilization of pastures.

Piracicaba: **Associaçâo Brasileira para Pesquisa da Potassa e do Fosfato**, 1986, p. 31-91.

MANRIQUE, L.A. Greenhouse crops: reviews. **Journal of Plant Nutrition**, v. 16, p. 2411-2477, 1993.

MAPA. **Tomato**, 2015. Available at:

<http://www.agricultura.mg.gov.br/images/documentos/perfil_tomate_mar_201 5[1].pdf> Accessed on: 02 Nov. 2016.

MARENCO, R. A.; LOPES, N. F. **Plant physiology**: photosynthesis, respiration, water relations and mineral nutrition. Viçosa, MG: Editora UFV, 2005, 451 p.

MARSCHNER, H.(1995) **Mineral Nutrition of Higher Plants**, 2 nd ed.

Academic Press, London.

MARTÎNEZ-ANDÙJAR, C.; GHANEM, M.E.; ALBACETE, A.; PÉREZ-ALFOCEA, F. Response to nitrate/ammonium nutrition of tomato (*Solanum lycopersicum* L.) plants overexpressing a prokaryotic NH4+ - dependent asparagine synthetase. **Journal of Plant Physiology**, v. 170, p. 676-678, 2013.

MARTINS, G.; CASTELLANE, P.D.; VOLPE, C.A. Influence of the greenhouse on climatic aspects during the rainy summer. **Horticultura brasileira**, Brasilia, v.12, n.2, p.131-135, 1994.

MEHMOOD, N.; AYUB, G.; ULLAH, I.; AHMAD, N.; NOOR, M.; KHAN, A.M.; AHMAD, S.; SAEED, A.; FARZANA. Response of tomato (*Lycopersicon esculentum* Mill.) cultivars to nitrogen levels. **Pure and Applied Biology**, v. 1, p. 63-67, 2012.

MENGEL, Y.; KIRKBY, E.A. **Principles of plant nutrition**. Bern: International

Potash Institute, 1987. 687p.

MORETTI. Protocols for Evaluating the Chemical and Physical Quality of Tomatoes. **Technical Communication 32**. Brasilia, DF, 2006.

NASCIMENTO, E.AL.; OLIVEIRA, L.E.M.; CASTRO, E.M.; DELU FILHO, N.; MESQUITA, A.C.; VIEIRA, C.V. Morphophysiological changes in leaves of coffee trees (*Coffea aràbica* L.) intercropped with rubber trees (*Hevea brasiliensis* Muell. Arg.) **Ciência Rural**, Santa Maria, v.36, n.3, p.852-857, 2005.

NASCIMENTO, J.; FAUSTINO, M. N. dos S.; MENESES, J. A. G.; SILVA, J. V.; SILVA, S. dos S.; CARVALHO, C. M. Initial growth of black string beans under different shading conditions and nitrogen fertilization.

IV WINOTEC - International Workshop on Technological Innovations in Irrigation. Fortaleza - CE, Brazil, 2012.

NASCIMENTO, T. B. do. **Quality of yellow passion fruit produced at different times in the south of Minas Gerais,** 1996. 56 f. Dissertation (Master's Degree in Plant Science) - Federal University of Lavras, Lavras, 1996.

NAVARRETE, M.; JEANNEQUIN, B.; SEBILLOTTE, M. Vigour of greenhouse tomato plants (Lycopersicon esculentum Mill.): Analysis of the criteria used by growers and search for objective criteria. **Journal of**

Horticultural Science, London, v.72, n.5, p.821-829, 1997

NOVAIS, R. F.; ALVAREZ, V. H. V.; BARROS, N. F.; FONTES, R. L. F.;

CANTARUTTI, R. B.; NEVES, J. C. L. **Fertility of the soil: Nitrogen**.

Viçosa-MG, p.375-470, 2007.

OAKS, A. (1994). Primary nitrogen assimilation in higher plants and its regulation. **Can. J. Bot**, 72: 739-750.

OAKS, A., HIREL, B. Nitrogen metabolism in roots. **Annual Review of Plant Physiology**, Palo Alto, v.36, p.345-365, 1985.

OBERLY, A.; KUSHAD, M.; MASIUNAS, J. Nitrogen and tillage effects on the fruit quality and yield of four tomato cultivars. **Journal of Vegetable Crop Production**, v.8, p.65-79, 2002.

OTONI, B. S. Characterization and production of two tomato hybrids (*Lycopersicon esculentum*) grown under different shading levels. 2010. 33f. **Dissertation** (Master's Degree in Plant Production in the Semi-Arid) - Postgraduate Course at the State University of Montes Claros. Janaùba, MG, 2010.

PACHECO, M. A. S. R.; FONSECA, Y. S. K.; DIAS, H, G. G.; CÂNDIDO, V. L. P.; PAZINATO, B. C. ; GALHARDO, R. C. Processamento artesanal do tomate. 2ª impresso. Campinas: Coordenadoria de Assistência Tècnica Integral, 1997. 30 p.

PAIVA, N. C. de. Vegetable production in a protected environment / Milton César de Paiva, Cuiabà: SEBRAE/MT, 1998. 85 p. (**Coleçâo Agroindùstria**, v. 18).

PAPADOPOULOS, A. P. Growing greenhouse tomatoes in soil and in soilless media. **Ottawa: Agriculture Canada Publication**, 1991. 79p.

PORTO, J. S. Nitrogen sources and doses in the production and quality of Silvety hybrid tomatoes. Dissertation (Master's Degree in Agronomy) - Postgraduate Course at the State University of Southwest Bahia. Vitória da Conquista, BA, 2013.

PASSAM, H.C.; KARAPANOS, I.C.; BEBELI, P.J.; SAVVAS, D. A review of recent research on tomato nutrition, breeding and post-harvest technology with reference to fruit quality. **The European Journal of Plant Science and Biotechnology**, v. 1, p. 1-21, 2007.

PERALTA, I.E.W.; SPOONER, D.M. Granule-bound starch synthetatse (GBSSI) gene phylogeny of wild tomatoes (*Solanum* L. section *Lycopersicon* (Mill) Wettst. Subsection *Lycopersicon*). **American Journal of Botany**, v.88, p.1888-1902, 2001.

PERALTA, I.E.; KNAPP, S.; SPOONER, D.M. Nomenclature for wild and cultivated tomatoes. **TGC Report**, v.56, p.6-12, 2006.

http://tgc.ifas.ufl.edu/vol56/html/vol56featr.htm (October 22, 2016).

PEZZOPANE, J.E.M; PEDRO JR, M.J.; ORTOLANI, A.A. Microclimatic changes caused by greenhouses with plastic cover. **Bragantia**, v.54, p. 419-425, 1995.

PIERRÔ, A. Good Taste. **Cultivar Hortaliças e Frutas**, n.14, p.10-12, 2002.

RADIN, B. ; BERGAMASCHI, H.; REISSER JUNIOR, C.; BARNI, N. A.; MATZENAUER, R. DIDONÉ, I. A. Efficiency of use of photosynthetically active radiation by the tomato crop in different environments **Pesquisa Agropecuària Brasileira**, Brasilia, Sep. 2003, v. 38, n. 9, p. 10171023.

RAJAPAKSE, N. C.; SHAHAK, Y. Light quality manipulation by horticulture industry. In: Whitelam, G. C.; HALLIDAY, K. J. **Light and Plant Development**, Leicester, 2007. p. 290-312.

RASMUSSEN, G.K. et al. The organic acid content of Valencia oranges from four locations in the United States. **Proceedings of the American Society Horticultural Science,** Alexandria, v.89, p.206-210, 1966.

REIS, L. S.; SOUZA, J. L.; AZEVEDO, C. A. V. de. Evapotranspiration and crop coefficient of kaki tomatoes grown in a protected environment**. Revista Brasileira de Engenharia Agricola e Ambiental,** v.13, p.289-296, 2009.

RIGUI, C. A.; BERNARDES, M. S. A. Availability of radiant energy in an agroforestry system with rubber trees: Bean productivity, **Bragantia**, v.67, p.533-540, 2008.

RITZINGER, R.; MANICA, I.; RIBOLDI, J. Effect of spacing and harvest time on the quality of yellow passion fruit. **Pesquisa Agropecuària Brasileira**, Brasilia, v.24, n.2, p.241-245. 1989.

ROCHA, R. C. Use of different shading screens in the protected cultivation of tomato plants. 2007. 90f. **Thesis** (Doctorate in Agronomy) - Postgraduate Course in Agronomy - Horticulture, Universidade Estadual Paulista, Faculdade de Ciências Agronômicas, Botucatu, 2007.

RODRIGUES, 2015. Protected agriculture: cooperation predicts technological advances in the protected cultivation of vegetables. Hortaliças em Revista.

EMBRAPA HORTALIÇAS. Year IV, n° 17, Sep-2015.

RONCHI, C. P.; FONTES P. C. R.; PEREIRA, P. R. G.; NUNES, J. C. S.; MARTINEZ, H. E. P. Nitrogen indices and tomato growth in soil and nutrient solution. Revista Ceres, v. 48, p.469-484, 2001.

RUDICH, J., GEIZENBERG, C., GERA, G., KALMAR, D., HOVEL, S. Drip irrigation of late seeded tomato for processing. **Acta Horticulturae**, n. 89, p.5968. 1979.

SAINJU, U.S.; DRIS, R.; SINGH, B. Mineral nutrition of tomato. **Food**

Agriculture and Environment, v. 1, p.. 176-183, 2003

SANTI, A. Tomato production in substrates and shading levels in Tangara da Serra/MT. Thesis (Doctorate in Tropical Agriculture) - Postgraduate Course in Tropical Agriculture. Cuiabâ. MT, 2014.

SANTOS, F.F.B. Breeding and selection of tomato hybrids for resistance to *Tomato yellow vein streak virus* (TOYVSV). 2009. **Dissertation** (Master's Degree in Tropical and Subtropical Agriculture) - Instituto Agronômico, Campinas, SP.

SCAIFE, A.; BAR-YOSEF, B. **Nutrient and fertilizer management in field grown vegetables.** Basel: International Potash Institute, 1995. 104p. (IPI.

Bulletin, 13).

SCALON, S. de P.Q. et al. Germination and growth of pitangueira (Eugenia uniflora L.) seedlings under shading conditions. **Rev. Bras. Frutic.,** Jaboticabal, v. 23, n. 3, 2001.

SCHWARZ K; RESENDE JTV; PRECZENHAK AP; PAULA JT; FARIA MV; DIAS DM. 2013. Agronomic performance and physicochemical quality of tomato hybrids in creeping cultivation. **Horticultura Brasileira** 31: 410-418.

SEGATTO, F.B.; BISOGNIN, D.A.; BENEDETTI, M.; COSTA, L.C.;

RAMPELOTTO, M.V.; NICOLOSO, F.T. Technique for studying the anatomy of potato leaf epidermis. **Revista Ciência Rural**, v.34, n.5, set-out, 2004, p.1597

SILVA, C.R.; VASCONCELOS, C.S.; SILVA, V.J.; SOUSA, L.B.; SANCHES, M.C. Growth of tomato seedlings with different shading screens. **Supplement**. Uberlândia, v.29, p.1415-1420, nov. 2013.

SILVA, J.B.C.; GIORDANO, L.B. Tomato for industrial processing. Brasilia: Embrapa Comunicação para Transferência de TecnologiaZEmbrapa Hortaliças, p.8-11, 2000.

SIMS W.L. History of tomato production for industry around the world. **Acta Horticulturae**, vol.100, p.25-26, 1980.

SJOSTROM, G.; ROSA, J. F. L. Study on the physical characteristics and chemical comparison of yellow passion fruit, *Passiflora edulis* f. *flavicarpa* Deg. grown in the municipality of Entre Rios, Bahia. In: CONGRESSO BRASILEIRO DE FRUTICULTURA, 4., 1977, Salvador. **Proceedings...** Caçador: Sociedade Brasileira de Fruticultura, 1978. p.265-73.

SOUZA, J.A.R.; MOREIRA, D.A. Evaluation of table tomato fruits produced with effluent from the preliminary treatment of swine wastewater. **Engenharia Ambiental**, v. 7, p. 152-165, 2010.

SPOONER, D.M.; HETTERSCHEID, W.L.A.; VAN DEN BERG, R.G.; BRANDENBURG, W. Plant nomenclature and taxonomy: an horticultural and agronomic perspective. **Horticultural Review**, v.28, p.1-60, 2003.

http://media.wiley.com/product_data/excerpt/22/04712154/0471215422.pdf. (November 15, 2016).

STRECK, N.A.; BURIOL, G.A.; ANDRIOLO, J.L.; SANDRI, M.A. Influence of plant density and dratic apical pruning on tomato productivity in a plastic greenhouse. **Pesquisa Agropecuâria Brasileira**, Brasilia, v.33, n.7, p.1105-1112, 1998.

SWIADER JM; MOORE A. 2002. SPADchlorophyll response to nitrogen fertilization and evaluation of nitrogen status in dryland and irrigated pumpkins. **Journal of Plant Nutrition** 25: 1089-1100.

TAIZ, Lincoln & ZEIGER, Eduardo. **Plant physiology**. 4ª edição, Porto Alegre, p.

317-339, 2009.

TROEH, F. R.; THOMPSON, L M. **Soils and soil fertility**. Sao Paulo, 718p., 2007.

WARNOCK, S.J. A review of taxonomy and phylogeny of genus *Lycopersicon.* **HortScience,** Alexandria, v.23, n.4, p.669-673, 1988.

WILLIAMS, J.W., SISTRUNK, W.A. Effect of cultivar, irrigation, etephon and harvest date on the yield and quality of processing tomatoes. **Journal of the American Society for Horticultural Science**, v. 104, n.4, p. 435-439. 1979.

WILLIAMS, L. E.; MILLER, A.J. Transporters responsible for the uptake and partitioning of nitrogen solutes. **Plant Molecular Biology**, v.52, p.659-688, 2001.

WITTWER, S.H.; CASTILLHA, N. Protected cultivation of horticultural crops wordwide. **HortTecnology**, Barking, v.5, n.5, p. 6-23, Jan. 1995.

YRISARRY, J.J.B., LOSADA, M.H.P., RINCON, A.R. del. Response of processing tomato to three different levels of water and nitrogen applications. **Acta Horticulturae**, n. 355, p. 149-156, 1993.

Printed by Books on Demand GmbH, Norderstedt / Germany